Global Ecology and Unequal Exchange

In modern society, we tend to have faith in technology. But is our concept of 'technology' itself a cultural illusion? This book challenges the idea that humanity as a whole is united in a common development towards increasingly efficient technologies. Instead it argues that modern technology implies a kind of global 'zero-sum game' involving uneven resource flows, which make it possible for wealthier parts of global society to save time and space at the expense of humans and environments in the poorer parts.

We tend to think of the functioning of machines as if it was detached from the social relations of exchange that make machines economically and physically possible (in some areas). But even the steam engine that was the core of the Industrial Revolution in England was indissolubly linked to slave labour and soil erosion in distant cotton plantations. And even as seemingly benign a technology as railways have historically saved time (and accessed space) primarily for those who can afford them, but at the expense of labour time and natural space lost for other social groups with less purchasing power. The existence of technology, in other words, is not a cornucopia signifying general human progress, but the unevenly distributed result of unequal resource transfers that the science of economics is not equipped to perceive. Technology is not simply a relation between humans and their natural environment, but more fundamentally a way of organizing global human society. From the very start it has been a global phenomenon, which has intertwined political, economic, and environmental histories in complex and inequitable ways. This book unravels these complex connections and rejects the widespread notion that technology will make the world sustainable. Instead it suggests a radical reform of money, which would be as useful for achieving sustainability as for avoiding financial breakdown.

The book brings together various perspectives from environmental and economic anthropology, ecological economics, political ecology, world-system analysis, fetishism theory, semiotics, environmental and economic history, and development theory. Its main contribution is a new understanding of technological development and concerns about global sustainability as questions of power and uneven distribution, ultimately deriving from the inherent logic of general-purpose money. It should be of interest to students and professionals with a background or current engagement in anthropology, sustainability studies, environmental history, economic history, or development studies.

Alf Hornborg is an anthropologist and Professor of Human Ecology at Lund University, Sweden.

Routledge studies in ecological economics

1 **Sustainability Networks**
Cognitive tools for expert collaboration in social-ecological systems
Janne Hukkinen

2 **Drivers of Environmental Change in Uplands**
Aletta Bonn, Tim Allot, Klaus Hubaceck and Jon Stewart

3 **Resilience, Reciprocity and Ecological Economics**
Northwest coast sustainability
Ronald L. Trosper

4 **Environment and Employment**
A reconciliation
Philip Lawn

5 **Philosophical Basics of Ecology and Economy**
Malte Faber and Reiner Manstetten

6 **Carbon Responsibility and Embodied Emissions**
Theory and measurement
João F. D. Rodrigues, Alexandra P. S. Marques and Tiago M. D. Domingos

7 **Environmental Social Accounting Matrices**
Theory and applications
Pablo Martínez de Anguita and John E. Wagner

8 **Greening the Economy**
Integrating economics and ecology to make effective change
Bob Williams

9 Sustainable Development
Capabilities, needs, and well-being
Edited by Felix Rauschmayer, Ines Omann and Johannes Frühmann

10 The Planet in 2050
The Lund discourse of the future
Edited by Jill Jäger and Sarah Cornell

11 Bioeconomics
Edited by Mauro Bonaiuti

12 Socioeconomic and Environmental Impacts on Agriculture in the New Europe
Post-Communist transition and accession to the European Union
S. Serban Scrieciu

13 Global Ecology and Unequal Exchange
Fetishism in a zero-sum world
Alf Hornborg

'Writing with clarity and analytical focus, Alf Hornborg takes on our most serious problems today: the degradation of the environment and the unequal distribution of wealth. Going beyond the usual stories about equal exchange, he shows how resources from labor to energy are unequally transferred across the globe to the detriment of the environment and equity: our narratives about the benefits of growth and efficiency hide the realities of human and material debasement. Drawing on anthropology, Hornborg persuasively argues that by distinguishing money's several purposes through issuing multiple currencies, and by adjusting our tax and subsidy policies, we may reduce our footprints and cope more sensibly.'

Stephen Gudeman, *University of Minnesota, USA*

'We view technology as akin to magic, creating value from nothing. Blending culture and ecological economics, Alf Hornborg shows that technical development is actually a zero-sum game. Technology transfers assets between countries, producing wealth in some regions while imposing labor and environmental costs elsewhere. All the while, money makes material resources invisible. Civilization, concludes Hornborg, has been built on the displacement of environmental damage. This book should be read by all persons concerned with technology, development, and trade.'

Joseph Tainter, *Utah State University, USA*

Global Ecology and Unequal Exchange

Fetishism in a zero-sum world

Alf Hornborg

Routledge
Taylor & Francis Group
LONDON AND NEW YORK

First published 2011
by Routledge
2 Park Square, Milton Park, Abingdon, Oxon OX14 4RN

Simultaneously published in the USA and Canada
by Routledge
711 Third Avenue, New York, NY 10017

Routledge is an imprint of the Taylor & Francis Group, an informa business

British Library Cataloguing in Publication Data
A catalogue record for this book is available from the British Library

Library of Congress Cataloging in Publication Data
Hornborg, Alf.
Global ecology and unequal exchange : fetishism in a zero-sum world / by Alf Hornborg.
p. cm.
Includes bibliographical references and index.
1. Environmental policy–Economic aspects. 2. Economic development–Technological innovations. 3. Economic policy. 4. Commodity fetishism. 5. Environmental sciences. 6. Human ecology. I. Title.
HC79.E5.H667 2011
333.7–dc22

2011005713

ISBN: 978-0-415-61486-3 (hbk)
ISBN: 978-0-203-80689-0 (ebk)

Typeset in Times
by Wearset Ltd, Boldon, Tyne and Wear

This book is dedicated to my three grandchildren Elias, Ella, and Henrik

Contents

Tables

Acknowledgements

For the research on which this book is based, I gratefully acknowledge financial support from The Swedish Research Council for Environment, Agriculture and Spatial Planning (FORMAS). In particular, I am grateful to FORMAS for supporting the Lund University Centre of Excellence for the Integration of Social and Natural Dimensions of Sustainability (LUCID), which has provided me with the opportunity to assemble this volume.

For permission to reproduce revised versions of texts previously published or in press elsewhere, I thank the following journals and publishers:

Chapter 1: This paper has been published in the *International Journal of Comparative Sociology* 50(3–4)237–62 (2009) by SAGE Publications Ltd., All rights reserved. © A minor part of it was also published in *Anthropology Today* 24(5):4–5 (2008).

Chapter 2: Parts of this text overlap with an article published by Taylor & Francis in the journal *Ethnos* 71(1):21–32 (2006).

Chapter 3: Parts of this text have appeared in my introduction to *Rethinking Environmental History* (Hornborg, McNeill and Martinez-Alier 2007) and in my chapter in the volume *Sustainable Development in a Globalized World*, edited by B. Hettne (Palgrave Macmillan, 2008).

Chapter 4: This version originally appeared in *REVIEW: A Journal of the Fernand Braudel Center*, XXXIII:2 (2010). A slightly shorter version has appeared electronically in *The Encyclopedia of Life Support Systems*, published by UNESCO (2010), and a minor part of it is published in *Handbook of World-Systems Analysis*, edited by S. Babones and C. Chase-Dunn (Routledge, 2011).

Chapter 5: Much of this text originally appeared in the Elsevier journal *Ecological Economics* 59(1):74–81 (2006) and was reprinted in *Rethinking Environmental History* (Hornborg, McNeill and Martinez-Alier 2007), published by AltaMira Press, and in *Recent Developments in Ecological Economics*, edited by J. Martinez-Alier and I. Röpke (Edward Elgar, 2008).

Chapter 6: This text was published in *The Austrian Journal of Development Studies* (March 2011). A part of it overlaps with an article published in the *Journal of Ecological Anthropology* 7:4–10 (2003).
Chapter 7: An earlier version of this text was published by the journal *Sign Systems Studies* 29(1):121–52 (2001).
Chapter 8: A part of this text overlaps with an article published in *Culture and Agriculture* 29(2):63–69 (2007).

Introduction

This book makes three assumptions about its readers. The first is that you are seriously concerned about what humanity (or at least a part of humanity) is doing to the rest of the biosphere, or 'the environment'. The second is that you are seriously concerned about global inequalities in the distribution of money, technology, and power. The third is that you are convinced that mainstream thinking about these matters is fundamentally flawed in the sense that it will not lead to the kind of changes that you are hoping for, summed up in visions of a sustainable development. If all these three assumptions apply in your case, then I hope you will be prepared to spend some time considering the perspectives offered in this book. Building on those concerns, it attempts to outline an alternative understanding of the connections between global environmental problems and global inequalities.

My own background is in cultural anthropology, so it is only to be expected that my struggles to rethink environmental issues, economics, and technology tend to take cultural aspects as a point of departure. The foremost hallmark of anthropology, I would argue, is the will to perceive and analyse the cultural dimension of all human thought and activity. But although much has been written under the banners of 'environmental anthropology', 'economic anthropology', and the 'anthropology of technology', it remains a demanding task to unravel how we modern humans tend to be constrained by our cultural preconceptions about material things such as the environment, the economy, and technology.

These difficulties, of course, go back to the tenacious separation of social and natural science. In the world of research, at least, it is obvious that those who seem to be the most concerned about the future of the biosphere (the natural scientists) generally have the bluntest analytical tools for understanding the causes of anthropogenic environmental degradation, while those who possess such tools (social scientists, including anthropologists) generally tend to be less concerned with the biosphere as an objective, biophysical reality. In having been trained to approach the environment with quotation marks, as a 'contested' or 'negotiated' cultural construction, anthropologists often seem at a loss when expected to say something about the *real* environment beyond human perceptions. Certainly, natural scientists need to realize that cultural systems such as language, money, and power are components of ecosystems, organizing significant aspects of their flows of matter and energy, but social scientists conversely need to realize that flows of matter and energy are fundamental to social systems, and need to be taken into account in any explanation of development and underdevelopment. Given this divide between social and natural science, the real dilemma is how we can learn to think in truly transdisciplinary ways about how material and semiotic processes are intertwined.[1]

So how could such transdisciplinary arguments, showing how material and cultural aspects interact, shed new light on the environment, the economy, and technology? We can begin by suggesting that the mainstream view of international trade, as promulgated in the science of economics, is itself largely a cultural perspective (cf. Gudeman 1986). Even today, and most obviously during the nineteenth century when the science of economics was emerging, it reflects the outlook of a particular category of people, positioned in economically affluent regions of the world. This view of international trade, which continues to dominate official thought and policy, is focused on the opportunities to generate monetary profits (somewhat euphemistically referred to as 'economic growth'), while more or less completely disregarding the inequitable material consequences of trade in terms of the state of the biophysical environment, human health, and disparities in the distribution of technological infrastructure. Such material parameters have largely been left out of the picture.

This is one reason why the concept of 'unequal exchange' is generally perceived by economists as irrelevant or incomprehensible. The assumption is that market prices guarantee fair trade. Once a commodity or service has been paid for, market ideology seems to assume, both sellers and buyers have every reason to be content. But this assumption has for a long time been challenged from various vantage points, ranging from the Marxist concern with the underpayment of labour to more recent concerns with the underpayment of natural resources. The fundamental difficulty faced by such challenges, however, has been how to define or even posit the existence of 'underpayment', if the only conceivable metric is money prices. If every market transaction is compensated with the money price that both parties have agreed to, what basis could there possibly be for suggesting that the exchange is unequal? This book tries to deal with this dilemma by arguing that monetary compensation is merely a surface phenomenon, an ideology of reciprocity, beyond which we can discern highly asymmetric flows of resources that can be measured in other metrics, such as labour time, hectares of land, energy, or material volume. Although largely invisible for the science of economics, such net transfers of physical resources are fundamental to the conspicuously uneven accumulation of technological infrastructure. To grasp this argument, as I emphasize particularly in Chapter 6, it is imperative to make an analytical distinction between the culturally defined domain of utility, consumer value, and market prices, on one hand, and the material domain of biophysical flows that are orchestrated by these market prices, on the other. This is the only way we can hope to persuade mainstream economists that there is something missing in their models, which can help us explain why global disparities in environmental degradation, human health, and technological development not only persist but even widen. Since every city from Uruk to Chicago (cf. Cronon 1991) is incontrovertible evidence of ecologically unequal exchange, the question for ecological economics cannot be whether such exchange is relevant to accumulation, but the extent to which it has been extended to the international arena.

In adopting such an alternative perspective on the economy, we would need to rethink much of what is today taken for granted in the dominant world-view. As I show in the first chapter, it would lead us to rethink not only economics and the notion of market prices, but also the idea that economic growth and technological development can be considered a cornucopia signifying general human progress. Global development gaps that we tend to perceive as reflecting stages in time can more clearly be recognized as inequalities in societal space. The technological capacity of a given population can be largely attributed to that population's position in a global system of resource flows, rather than to some innate predilection to 'invent'. Finally, if global society is as founded on contradictory interests as this analysis suggests, the prospects of achieving sustainable development through voluntary frugality and consensus, rather than conflict, seem slim. For this reason, the topic of socio-ecological breakdown, or 'collapse', is a recurrent consideration in several chapters.

An indispensable component in my argument is the concept of 'machine fetishism', which also recurs throughout the book. In Chapter 2, I show how the Marxian concept of fetishism can be extended from our understanding of money and commodities to explain how we tend to be deluded by modern technology. All three categories of objects (money, commodities, and machines) are fetishes in the sense that they mystify unequal relations of exchange by being attributed autonomous agency or productivity. Again, a cultural perspective is fundamental to understanding how the inequality of global exchange has been kept out of view. The mainstream interpretation of modern technology, of course, is that it is an index of human progress over time, even as a gift to humanity from the wealthier nations of the world. The zero-sum game perspective pursued in this book instead argues that modern technology is largely an index of accumulation, rather than ingenuity, and that its capacity to locally save time and space occurs at the expense of (human) time and (natural) space lost elsewhere in the world. This is demonstrated, for instance, in Chapter 5, where I calculate the extent to which the Industrial Revolution in England was founded on what I call 'time–space appropriation', a concept which combines the Marxist focus on the unequal exchange of labour time with more recent, ecological concerns with the unequal exchange of embodied land.

In Chapters 3 and 4, I show how these perspectives should change our ways of writing history, particularly environmental and economic history, but ultimately also the history of technology. Not only must Europe and the 'West' be dethroned as intrinsically generative of economic growth, modern technology, and civilization, but these phenomena must in themselves be recognized as contingent on specific global constellations of asymmetric resource flows and power relations. In other words, not only was the 'rise of the West' a geographical coincidence of world history – the location of Europe as middleman between the Old and New Worlds (cf. Blaut 1993, 2000; Frank 1998; Pomeranz 2000) – but its economic, technological, and military *means* of expansion, generally viewed as European 'inventions' and as contributions to the rest of humanity, were products of global conjunctures and processes of accumulation that coalesced after the articulation of the Old and New Worlds. The very existence of industrial technology has thus from the very start been a global phenomenon, which has intertwined political, socio-economic, and environmental histories in complex and inequitable ways. Technological rationality is never disconnected from issues of global resource distribution. If historical hindsight helps to clarify this generally neglected fact, the next challenge must be to spell out its ramifications for our perceptions of economic growth and technological progress *today*. We need to understand that technology is not simply a relation between humans and their natural environment, but more fundamentally a way of organizing global human society.

The final two chapters take a closer look at the (cultural) phenomenon of *money*. In Chapter 7, I apply what I call an *ecosemiotic* perspective on money, comparing monetary flows with other kinds of signs flowing through human societies and ecosystems. Using examples of human–environmental relations in Amazonia, I show how monetary sign systems differ from sensory and linguistic sign systems, and have very different consequences for the societies and ecosystems in which they operate. In Chapter 8, finally, I argue that what economic anthropologists call general-purpose money is intrinsically destructive of ecosystems and that our thinking about alternative, more sustainable economic systems ought to take this into account. This chapter discusses various ideas about how to intervene in the destructive logic of our current economic system and suggests that our best bet is to reorganize the way money operates. It actually sketches a tangible policy for such a reform, which could be as useful for achieving sustainability as for avoiding financial breakdown. Although beyond the scope of this book, the current worries about financial crises have a direct connection to its main argument, as these crises illustrate what can happen when monetary and material assets become too dissociated from each other, while the treadmill logic of expanding consumption continues to drive the economy. As such financial problems increase over the next few decades, it is reasonable to expect the ideas in this book to sound more compelling, and its policy suggestions more realistic.

Alf Hornborg

1 Zero-sum world

How to think about ecologically unequal exchange

Researchers and policy-makers addressing sustainability in the early twenty-first century face a challenge so formidable that they may prove incapable of dealing with it, in which case our current capitalist civilization may well share the fate of ancient Rome and similar historical instances of socio-ecological collapse (Tainter 1988). This alarmist introduction is meant to underscore the urgency of the analytical task that I attempt to outline in this book. The currently globalizing connections through market exchange and technologies of trade and communication are widely celebrated as a road to a more integrated, prosperous, and even egalitarian future world, yet there is overwhelming evidence that precisely these connections continue to generate devastating ecological deterioration and increasingly severe inequalities within and between nations (United Nations Development Program 1998; Millennium Ecosystem Assessment 2005). Almost seven billion human beings are currently implicated in a global system that seems inexorably to bring us all closer and closer to socio-ecological collapse. There is nothing inevitable about this process, many of us are aware of its fundamental direction, yet we seem quite unable to halt it.

This incapacity to evade catastrophe has two basic aspects that are intricately interrelated. One is that our way of thinking and talking about the world prevents us from grasping or at least efficaciously questioning the mechanisms propelling this development. The other is that there are extremely powerful interests at stake. We are not all sitting in the same boat, as the metaphor goes. We are sitting in at least two different boats, but one is pulling us all toward disaster. There are definitely powerful social groups who have very much to gain – at least within the anticipated timeframe of their own lifetimes – from the current organization of global society. As many social scientists have shown, it is precisely these groups who tend to exert a primary influence over the way social processes are defined – and even questioned. The language devised to manage socio-ecological 'problems' viewed through such system-serving lenses will naturally constrain our capacity to actually 'solve' problems in the sense of changing the direction of societal development, which may well require fundamentally reorganizing social institutions. The language of policy and management thus tends to avoid questions of power, conflicts, and inequalities. Although conspicuously present – and increasingly problematic – in global human society, such issues are rarely identified as problems to be solved. There is rather a pervasive assumption of consensus with regard to appropriate policy and management.

A crucial challenge for social sciences struggling with these issues is that the everyday assumptions about the world in which we are all suspended, and which are generally described for us in terms of flows of money and information, have very tangible material properties and consequences. These material aspects of global society are widely ignored in social science, in part because they implicate knowledge and methodologies generally reserved for the natural sciences. Nor can they be fully grasped by the natural scientists themselves, simply because these researchers generally have a poor understanding of society. Yet the logic of these material aspects of society – what are increasingly referred to as 'socio-ecological' systems – urgently needs to be understood (Berkes and Folke 1998; Hornborg and Crumley 2007). But even here, in contemporary attempts to transcend the academic distinction between social and natural sciences, there is a clear divergence between perspectives emphasizing, respectively, consensus and conflict. In this book, I will take power, contradiction, and 'capital accumulation by dispossession' (Harvey 2003) as a point of departure for understanding the disastrous course of current socio-ecological processes, but I will also briefly demonstrate why the hegemonic interpretations and policies that instead assume consensus (e.g. the functionalist discourse on 'resilience') may be misguided.

The chapter is divided into three parts. The first discusses how a population's perceptions of technology, economy, and ecology are conditioned by its position within global systems of resource flows, and how mainstream modern perceptions of 'development' can be viewed as a cultural illusion confusing a privileged position in social space with an advanced position in historical time. The second part traces some lineages of critical thinking on environmental load displacement and ecologically unequal exchange, arguing that such acknowledgement of a global environmental 'zero-sum game' is essential to recognizing the extent to which cornucopian perceptions of 'development' indeed represent an illusion. The third part, finally, offers some examples of how the rising global anticipation of socio-ecological contradiction and disaster is being ideologically disarmed by the rhetoric on 'sustainability' and 'resilience'.

Machine fetishism: technology/economy/ecology as culture

In the mainstream language of policy for sustainable development, the words 'technology', 'economy', and 'ecology' are used in an unreflective, matter-of-fact way that suggests bounded categories of reality given once and for all and exempt from critical scrutiny. This is the language of positivism and simple empiricism, the diametrical antithesis to those traditions in social research that emphasize a deeper and second look at the surfaces of the world that present themselves to our senses. In this latter tradition I would include what David Harvey (1996) calls 'dialectics', but also the whole thrust of 'deconstruction' and 'defamiliarization' (Marcus and Fischer 1986) that has characterized so much of the work in humanities and some of the social sciences in recent decades. Researchers from these traditions will find it easier to digest what I am now going to propose, namely that our notions of 'technology', 'economy', and 'ecology' are *cultural* categories that train us to think about our socio-ecological realities in particular ways. These three categories represent overlapping phenomena, the analytical separation of which diminishes our chances of grasping the totality of which each gives a glimpse. The three concepts represent distinct and extremely influential fields of research lodged in separate academic faculties, yet each can simultaneously be used as a point of departure for extensive anthropological reflection on how mainstream thought is culturally constituted (e.g. Latour 1993; Sahlins 1976; Godelier 1986; Gudeman 1986; Croll and Parkin 1992; Descola and Pálsson 1996; Ellen and Fukui 1996; Ingold 2000; Hornborg 2001b). 'Technology' can thus be understood as referring to combustion engines and the development of new agrofuels, or it can be understood as a realm of fetishism, magic, and ritual. 'Economy' can be represented as dealing with market institutions and the measurement of GDP, or it can be thought of as concerned with ideology, permutations of reciprocity, and the driving forces of consumption. Finally, 'ecology' can be perceived as a biophysical domain of natural processes uncontaminated by human ideas and relations, or as the material-cum-relational substrate enveloping – and implicated in – all human life (including technology and economy).

David Harvey (1996) articulates the important but difficult ambition to bridge the divide between local particularities of experience, on one hand, and universalizing understandings of global socio-ecological processes, on the other. Much of the contemporary work in humanities and social sciences tends to focus on the former, to the exclusion of the latter. In this chapter I will have very little to say on local particularities of experience, but I agree with Harvey's conclusion that social science can and should try to account for how they are recursively related to global socio-ecological processes. A promising approach is to focus, as he does, on *money* as a social and cultural institution that generates 'space-time' as simultaneously an objective, political-ecological framework, and a subjective experience (e.g. of 'time–space compression'). Money is the very vehicle by which ideas about reciprocity and relations of exchange are translated into material processes capable of transforming not only human societies and technologies, but the entire biosphere (Hornborg 2001b). In looking at how different *kinds* of money can generate different kinds of material processes (or kinds of 'space-time'), we come closer to an understanding of what is required for us to actually make progress in our pursuit of sustainability.

The main argument regarding 'machine fetishism' is that the modern concept of 'technology' is a cultural category (Hornborg 1992, 2001b). It refers to what is technically feasible to achieve at a given time and place, but remains largely oblivious to the extent to which a local increase in technological capacity is a matter of shifting resources from one social category to another within global society. The notion of 'fetishism' can be applied so as to suggest that the apparent generative capacity of machine technology is an instance of how the attribution of autonomous productivity to material artefacts can serve to conceal unequal relations of exchange. The unequal exchange underlying machine technology can only be revealed by exposing, beyond the monetary price tags reified by conventional market ideology, material asymmetries in the net flows of biophysical resources gauged in terms of alternative metrics such as energy, matter, embodied land (ecological footprints), or embodied labour. The mechanical 'power' of the machine is thus an expression of the economic and ideological 'power' through which it is sustained. Ultimately, what keeps our machines running are global terms of trade.

The prospect of peaking oil extraction presently prompts us to rethink processes of development and decline in the world system. Rather than simply revive Malthusian concerns over the dismal destiny of humankind as a whole, we need to approach the popular notion of 'cheap energy' as an experience situated in societal space as well as in historical time. Energy has been perceived as 'cheap' only within core segments of world society, whose ideology of progress and development has tended to construe contemporary global inequalities as representing different stages in time. Draft animals and wood fuel are here often perceived as elements of the past, yet remain an everyday reality for significant parts of the world's population. Conversely, fossil-fuel technology is conceived as a 'now' rather than a 'here'. For many post-Soviet farmers, we should recall, the age of the machine is already a thing of the past. The machine is an index of purchasing power and a specific form of capital accumulation that is as mystified and fetishized as any other power strategy in history. As we begin to anticipate its demise, we might reflect on the fact that the war in Iraq and global climate change are opposite sides of the same coin. The structural problem of fossil-fuelled capitalism is to maintain imports of energy (e.g. oil) and exports of entropy (material disorder, e.g. in the form of carbon dioxide), two imperatives of 'development' that are both increasingly difficult to sustain.

There seems to be a growing expectation, at least in North America and Europe, that the age of fossil fuels is approaching an end. This anticipation of peaking oil and a post-petroleum era may at first glance seem a straightforward, 'practical' problem of technology and resource management, but it is in fact a condition that must be understood through the lens of cultural analysis, particularly a cultural analysis of power within the capitalist world system.

Let us begin by recalling that all technological systems are embedded in cultural – and political – systems of significance that tend to remain invisible (because self-evident) to the users of these technologies. Technologies are never 'merely' material strategies for getting certain kinds of work done; they also tend to embody tacit assumptions about their own rationality and efficiency. In other words, significant aspects of the functioning of technological systems rely on *beliefs* about their efficacy. Many anthropologists have thus already accepted that the boundary between technology and magic is difficult to draw. It is generally not difficult for us to imagine, for instance, how the ancient inhabitants of Easter Island found it imperative to struggle with those huge stone statues because they were perceived as *essential* for the implementation of some practical task, as understood through the local cosmology. We can similarly rest assured that the temple pyramids of the Maya and the sun rituals of the Inca – in this sense – should be regarded as technologies. To the extent that social life progressed, by and large, as these pre-modern peoples hoped, they no doubt perceived their respective technologies as efficient in relation to the tasks they were assumed to fulfil. Early sixteenth-century Andean harvests of corn, for instance, testified to the efficacy of the Inca emperor's sacrifices and ritual communication with his father the Sun.

But can our modern fossil-fuel technology really be compared to these exotic practices and cosmologies? Isn't the crucial difference that our machines actually *work*, whereas Inca ritual was just mystification? I don't think it is as simple as that. In both cases – the divine Inca emperor and our modern machinery – a particular material entity is perceived as productive or generative *in itself*: a cornucopia. In both cases, also, it is possible to demonstrate that productivity is the result not of properties intrinsic to either entity – the body of the Inca emperor or the design of the machine

– but of the societal flows of resources that reproduce, and are reproduced by, these entities. Without the asymmetric exchanges with his many subordinates at different levels, the Inca emperor would not have been able to fill his warehouses with the stores of food, textiles, and other treasures that so impressed the Spaniards. Without those exchanges, he would never have appeared a generous cornucopia. And today, without the asymmetric exchanges of high-quality energy on the world market, our machines would achieve absolutely nothing. As many post-Soviet farmers have experienced, when there is no longer any diesel in the tractor, it is just an assemblage of scrap metal. Again, what ultimately keep the machines running are global terms of trade.

Most of us are probably prepared to accept as a fact that resource flows within the Inca empire were asymmetric, or unequal. When the emperor's subordinates laboured on his fields in exchange for ceremonially served corn beer, it is obvious that their labour yielded more corn for his warehouses than they were able to consume in the form of beer, however great was their thirst. It is much more difficult to argue that the maintenance of modern machinery – as much as the maintenance of the Inca court – relies on unequal exchange. It is more difficult because *we* are now the ones inside the cultural bubble, so to speak. Much as the prevailing cosmology in sixteenth-century Peru probably made it difficult to question the divinity and productivity of the emperor, we are today suspended in a web of significance that makes it difficult to see – or at least to *say* – that industrial machinery relies on unequal exchange of resources in the world system. The currently hegemonic cosmology – known as economics – trains us to think that voluntary market transactions are by definition equal and fair. Of course, when gauged in terms of monetary price, they must be. But the cultural bubble of neoliberal economics excludes all those other possible measures of exchange – such as energy, materials, hectares, labour time – with which it is fairly easy to show that world trade is indeed highly unequal. The concept of 'market prices' thus performs an ideological function similar to the Inca concept of *minka*, that is, the ceremonial mobilization of labour in which the land-owning host was represented as generous, and the toil of the participants as reciprocation.

Maurice Godelier (1986) has argued that unequal exchange in human societies tends to present itself as a reciprocal exchange of services, and that economic anthropology should try to unravel the various ways in which this is done. Fundamental to these political arrangements is that the exploited are led to believe that they should be grateful to the exploiters. Inca ritual seems an obvious case, but are we now prepared to see that Godelier's observation is equally applicable to our own society and the entire industrial world order? Ever since the first major 'oil crisis' over thirty years ago, the dominant enthusiasm over development and growth has been accompanied by the uneasy realization that the success of our techno-economic cornucopia seems to hinge on the world market price of oil. The least disturbing implication that might be drawn is that there is a 'correct' price of oil, which is defined by the rate at which industrial economies can continue to export their commodities in exchange for ever greater quantities of oil. A more disturbing conclusion would be that there is no 'correct' price of oil, only a more or less profitable price, from the point of view of industrial economies. It would imply that when that price – that is, that profitable rate of exchange between industrial products and the resources that go into their production – is no longer tenable, our machines will grind to a halt.

After two centuries of living with machines propelled by fossil fuels, most of us are highly committed to the notion of ever more powerful technologies. The thought that the high-energy age of fossil fuels may turn out to be an historical parenthesis is generally dismissed as ridiculous. Yet we should be open to that possibility. The age of fossil fuels has not just been a period in time, but a condition situated in socio-political space. It has provided a minority of the world's population with an unprecedented source of *power* – in both a thermodynamic and a political sense. But we are now beginning to realize that the combustion of fossil fuels has represented an *illusory* emancipation from land. This illusory emancipation has two aspects. First, it has seemed to enable us to transcend the constraints of limited land area and soil fertility that so preoccupied the Physiocrats and other economic schools of thought prior to the Industrial Revolution. Second, it has until quite recently kept us largely ignorant about the negative consequences of burning fossil fuels for the long-term productivity of the biosphere as a source of human livelihood. For two and a half centuries, the more affluent parts of the world's population have been building a technology based on solar energy accumulated on the surface of the Earth a very long time ago. Each year, we have been dissipating energy representing *millions* of years of ecological production over significant parts of the Earth's surface. In other words, we have relied on acreages of the *past*. What the contemporary scramble for so-called biofuels (such as ethanol) really represents is our determination to try to sustain that same technology on the capacity of *presently* available land to accumulate solar energy. There seems to be a general confidence that it can be done. It is just a matter of getting the technology right. But what if it can't?

Generally speaking, social scientists will probably not get too involved in discussions about ethanol with all those engineers, agronomists, and economists who are committed to keeping the global technomass going by feeding it with corn or sugarcane. But we can listen attentively to the debate. We are told, for instance, that the conditions of people harvesting sugarcane for ethanol production in Brazil are appalling. We are told that ethanol production might in fact generate more greenhouse gases than the combustion of fossil fuels. We are told that it will accelerate tropical deforestation and loss of biodiversity. We are told that it will probably yield less horsepower per hectare than just simply growing fodder for horses. And what undoubtedly worries us the most, we are told that it is making food more expensive and contributing to malnutrition and starvation among the global poor. None of this should really come as a surprise. The technomass accumulated over two and a half centuries of fossil-fuelled capitalism is now competing with human and other biomass for living-space on this planet. The human agents committed to keeping this technomass growing have pursued various strategies for doing so, including military intervention in oil-producing countries, refusal to accept restraints on carbon dioxide emissions, and the appropriation of vast land areas for the production of alternative fuels.

Anthropologists can show how all of this can be understood in terms of a cultural analysis of power. Like the statues of Easter Island or the temple pyramids of the Maya, our machine technology has become fetishized to the point where it must be maintained at all cost, even at the cost of the land that can support us. And as we coax rural people in Brazil and Mozambique to devote their land and labour to support our technomass, the economists are seriously proposing that those rural people in the South are the ones who should be grateful – for the opportunity to 'develop'. Perhaps, at some point in the future, this will appear as absurd as it now would appear to us, if someone in sixteenth-century Peru had tried to persuade the peasants that one day, in the future, they would all have access to warehouses equal in size to those of the Inca emperor.

If we do accept that human understandings of socio-ecological processes are cultural (and necessarily also ideological, i.e. power-serving) constructions, we are in a better position to understand the European history of economic ideas as geared to the technological history of capital accumulation. Conventional economic science recognizes three significant 'factors of production', namely land, labour, and capital. Different degrees of significance have over time been attributed to these factors as sources of productivity or 'value' (cf. Gudeman 1986). In the agrarian world of the eighteenth-century Physiocrats, land was the ultimate source of value. In the nineteenth-century world of industrial factories, Adam Smith, David Ricardo, and Karl Marx championed labour as the main generative force. It could be argued that the twentieth century was dominated by the conviction that access to capital was the true bottleneck, as reflected, for instance, in the discourse on 'development'. As we proceed into the twenty-first century, there are signs that we may be returning to land as the key resource, as witnessed by the current scramble for vast acreages in Africa, Latin America, and Russia. As seems constantly to be the case in capitalist development (Harvey 2003), even this strategy for accumulation operates according to a principle of 'dispossession'. Accumulation by dispossession is the underlying logic of the unsustainable socio-ecological processes to which we shall now turn. Before doing so, however, we should pause to reflect on the implications of the above relativization of theories of 'value', including the Marxian labour theory of value and various versions of energy theories of value. Rather than ascribe to labour or energy uniquely generative powers, it should suffice to say that, for a successful capitalist, the price of labour (or energy) is significantly cheaper than the price of its products. The advantage with this formulation is that it is equally applicable to other indispensable inputs such as raw materials.

The metabolic rift: environmental load displacement and ecologically unequal exchange

For centuries, there has been a widespread intuition in both capitalist core nations and more peripheral areas that the economic and technological expansion of the former occurs at the expense of the latter. To simplify, we can call this world-view a 'zero-sum game' model of development, as opposed to the 'cornucopia' model, which instead proposes that capitalist growth in core areas is of benefit to their peripheries. Although morally compelling and widespread, the zero-sum game intuition has not become a part of mainstream thinking about development. Various kinds of discursive filters have resisted its incorporation in those versions of social science that are prominent in formulating policy and official political rhetoric. It is not difficult to understand this discursive resistance to moral qualms about development. In the core nations, politicians would be unwise to suggest that the average living standard in the country is unjustly high from a global perspective. In peripheral nations (the so-called 'developing' or 'less developed' nations), attempts by individual politicians to challenge global inequities and power structures have backfired in various ways, for instance through military interventions sponsored by the core, loss of economic benefits and support linked to established trade patterns, and the inability to offer a credible and attractive political programme to the electorate (cf. Hettne 1990). A fundamental problem is the inclination in both core and periphery to define 'progress' in terms of economic growth and technological advances (Norgaard 1994). A spatially restricted process of capital accumulation is thus presented as a temporal difference, i.e. the highly desirable future of all nations, and any politician offering anything else than such progress is not likely to have a future in politics.

Academics are widely assumed to conduct their research in accordance with established ideals of objectivity, integrity, and intellectual honesty. Yet most social scientists find themselves pursuing research agendas articulated by the politicians who ultimately control their funding, and subtle processes of selection tend to ensure that key positions of academic authority are not occupied by individuals with anything subversive to say about growth or development. Nevertheless, a minority of academics – today often referred to as 'activist scholars' – have been unable to suppress the 'zero-sum game' intuition and instead pursued various research strategies to substantiate it. I am not referring here to dissidence in general terms, which would include the entire Marxist tradition, but to that specific variant of dissidence that seeks to reveal how the accumulation of money and technology in core areas of the world system occurs at the expense of the natural resources, environment, and health of their peripheries. This kind of thinking can be given many labels. Perhaps the most encompassing is 'political ecology' (e.g. Peet and Watts 1996; Bryant and Bailey 1997; Low and Gleeson 1998; Martinez-Alier 2002; Paulson and Gezon 2005), but underneath this umbrella we can identify a number of more specific approaches, theories, and methodologies, such as 'environmental justice' (Harvey 1996), 'ecologically unequal exchange' (Bunker 1985; Hornborg 1998b; Rice 2007; Jorgenson and Clark 2009), 'ecological footprint analysis' (Wackernagel and Rees 1996), and so on. Common to all these approaches are concerns with the uneven deterioration of the natural environment, and with the relations of power that generate and maintain unevenly distributed environmental impacts in global society. There is also a concern, dating back to the earliest days of the Industrial Revolution, that modern forms of resource management, e.g. in industrial agriculture, are not at all as 'efficient' as the economists would have us believe, but instead wasteful and unsustainable. It is ironic, for instance, that the many observations on the negative 'energy return on (energy) investment' (EROI) in industrial agriculture can now be directed at the production of energy itself (Pimentel 1991, 2001; Patzek and Pimentel 2005; Pimentel *et al.* 2007). Joined together, these two concerns suggest that the wastefulness and unsustainability of industrial resource management is made possible by displacing environmental impacts to other areas, populations, or social categories. This is the central contention of what I have called the 'zero-sum game' understanding of development. It runs counter to the thrust of most social policy and discourse in industrial capitalism, yet has preoccupied a long line of analysts over the past two centuries.

The early history of this lineage of dissidents has been traced by Joan Martinez-Alier (1987). He reviews a long line of sceptical anti-modernists, many of whom were concerned with unsustainable flows of energy and nutrients in agriculture. These analysts, he concludes, were precursors to what is now known as 'ecological economics'. It is noteworthy that many of these early dissidents – e.g. Josef Popper-Lynkeus (1838–1921), Karl Ballod-Atlanticus (1864–1933), Sergei Podolinsky (1850–1891), and Frederick Soddy (1877–1956) – were heavily influenced by Karl Marx (1818–1883). Other early sources of inspiration include John Ruskin (1819–1900), Wilhelm Ostwald (1853–1932), Joseph Henry (1797–1878), and Thorstein Veblen (1857–1929). The influence of these eight men was in turn extended to their disciples, such as Otto Neurath (1882–1945), Vladimir Vernadsky (1863–1945), Eduard Sacher (1834–1903), Patrick Geddes (1854–1932), Henry Adams (1838–1918), Alfred Lotka (1880–1949), and Alexander Bogdanov (1873–1928). Among their students, in turn, we find names like Nicholas Georgescu-Roegen (1906–1994), Lewis Mumford (1895–1990), M. King Hubbert (1903–1989), and Leslie White (1900–1975), all prominent mid-twentieth-century critics of industrial civilization. Vernadsky's students Hutchinson and Lindeman became tutors for ecologists such as Ramon Margalef and the brothers Eugene and Howard Odum, and his student Georgescu-Roegen became the father of ecological economics, tutoring Herman Daly, Kenneth Boulding, and Garrett Hardin. White is recognized as one of the founders of ecological anthropology, and counted, for example, Marshall Sahlins and Elman Service among his students.

This rather restricted number of critics of industrial capitalism has had a decisive influence on the development of theories and methods for understanding unsustainable and inequitable use of natural resources. Although rarely acknowledged, many of these approaches can trace a direct genealogy back to Karl Marx. Thus, important parts of the intellectual ancestry of what is now known as ecological economics can be traced to Marx's strong concerns about the 'metabolic rift' (Foster 2000; Foster and Burkett 2004; Foster and Clark 2004; Burkett and Foster 2006). Marx was very much concerned with the asymmetric exchange of nutrients and other material resources between town and countryside in nineteenth-century Europe, which among other things resulted in the impoverishment of rural soils and the accumulation of garbage and sewage in urban areas. The deterioration of European soils prompted capitalist entrepreneurs to exploit phosphates in Oceania and deposits of guano along the west coast of South America, and to develop artificial fertilizers requiring significant inputs of energy. Marx's crucial observation can today be extended to the metabolism of the entire world system (Foster *et al.* 2010), some parts of which are ecologically impoverished while other parts are smothered with garbage, air and water pollution, and other forms of material overload. The pattern is quite simple, if not self-obvious. When too much biomass, nutrients, water, or other natural resources are removed, the result is loss of biodiversity, topsoil, fish stocks, or other vital assets (cf. Bunker 2007). Conversely, when too much matter and energy use is concentrated in an area, it may suffer from smog, acidification, eutrophication, accumulation of heavy metals, and problems with the disposal of solid waste. In fact, even the global logic of carbon dioxide emissions and climate change can be understood within this theoretical framework (Clark and York 2005).

As we turn to the work of modern researchers in this tradition, I will briefly show how they fit into the genealogy of development dissidence sketched above. We have already mentioned Nicholas Georgescu-Roegen and Howard T. Odum as heirs to the criticism of industrial energy flows going back to the mid-nineteenth century. Georgescu-Roegen (1971) brought economics and thermodynamics together by showing how production processes necessarily and inexorably increase total *entropy*, i.e. material and energetic disorder. Odum (1971) was influenced by the so-called Technocrat movement led by the above-mentioned M. King Hubbert. He offered an energy theory of value similar to that of the Technocrats, suggesting that *emergy* ('energy memory') – i.e. the amount of solar energy spent to produce a given commodity or organism – is an adequate measure of economic utility (Odum 1988). In a paper co-authored with J. E. Arding, Odum used this notion of 'emergy' to argue that trade between nations at different levels of technological development tended to be unequal in terms of net flows of energy (Odum and Arding 1991). Odum's concepts of 'emergy' and 'embodied energy' are clearly similar to Marx's concept of embodied labour, and both shared the problematic conviction that such past investments (of energy or labour) should be a measure of value.

Odum's attempts to ground economic increases in (exchange) value in thermodynamics can be criticized from the perspective of both economics and thermodynamics (Hornborg 1998b). In physical terms, as Georgescu-Roegen showed, increases in value in a production process correspond to a *decrease* in total energy quality (*negentropy*, or 'negative entropy', closely related to *exergy*). Accumulation of 'emergy' or 'energy memory' thus represents a dissipation of exergy. The concepts of emergy and exergy, in other words, suggest inverse ways of describing processes of energy transformation. Although a common pitfall, ever since the days of Podolinsky, has been to seek a thermodynamic foundation for an objective measure of value, more 'real' than price (cf. Hornborg 1992), such arguments should be abandoned. The driving forces of consumption are cultural, both in general terms and in specifics (Sahlins 1976). As I argue more elaborately in Chapter 6, 'value' must pertain to what human beings perceive, rather than to physics. In fact, it is only by keeping human valuation and physical properties analytically separate that we can reveal the destructive logic of capitalist processes: the more people are willing to pay for a particular product, the faster will be the dissipation of resources required to produce it. Because of the universal interchangeability of all things on the capitalist world market, the accelerating dissipation of resources will be rewarded with increasing amounts of resources to dissipate.

At about the same time that Georgescu-Roegen and Odum published their most widely cited books, two other scientific approaches to global inequalities were emerging. One, explicitly lodged in the Marxist tradition, was concerned with revealing asymmetries in international exchange, e.g. through unequal net flows of embodied labour (Emmanuel 1972) and structural relations of dependency between peripheries and cores, or satellites and metropoles (Frank 1967). This framework, itself building on earlier work on structural inequalities in Latin America (Prebisch 1950), became fundamental to what today is known as world-system analysis (Wallerstein 1974–1989). The other, pioneered by Georg Borgström (1965) and William Catton (1980), focused on the international appropriation of what we might call 'embodied land', i.e. the consumption of resources requiring land surfaces in excess of what is nationally available. Both these approaches have stimulated interesting new research over the past two decades. By combining the world-system perspective of Frank and Wallerstein with Emmanuel's notion of unequal exchange and Odum's notion of energy value, Stephen Bunker (1985) assembled the first formulation of a concept of ecologically unequal exchange. A few years later, Mathis Wackernagel and William Rees (1996) popularized Borgström's and Catton's notion of 'ghost acreages' into the highly influential concept of an 'ecological footprint', i.e. the per capita quantity of eco-productive land surface required to sustain a given level of resource consumption. Several researchers have found this measure useful in analysing uneven flows of international trade, as flows of commodities can be converted into flows of 'embodied footprints', i.e. embodied land (Andersson and Lindroth 2001; Jorgenson 2003; York *et al.* 2003; Jorgenson and Rice 2005, 2007; Andersson 2010; Nordlund 2010).

The Marxian concern with unequal exchange and the metabolic rift can also be traced to three additional modern scholars with a wide sphere of influence in political ecology and sustainability studies. I will first mention Joan Martinez-Alier, who compiled the above-mentioned genealogy of ecological economics, but whose later publications (e.g. 2002, 2007, 2010) have addressed current 'ecological distribution conflicts', particularly in the so-called South. His (2002) conclusions about the North's 'ecological debt' to the South have recently been confirmed by a study published by the US National Academy of Sciences (Srinivasan *et al.* 2008). Martinez-Alier's ambition seems to be to create a synthesis of environmentalism and Marxism, in the vein of what is often referred to as 'ecosocialism' or 'ecological Marxism'. This also applies to John Bellamy Foster, who has published extensively on the relation between Marxism and ecological theory (e.g. Foster 2000; Foster and Burkett 2008; Foster *et al.* 2010). However, Foster and Martinez-Alier have been in disagreement about the history of this relationship (Foster and Burkett 2004; Burkett and Foster 2006, 2008). Paul Burkett and John Bellamy Foster argue that Marx and Engels were much better acquainted with ecological theory than Martinez-Alier has suggested, and that their understanding of industrial capitalism was not only compatible with, but strongly informed by, the laws of thermodynamics. In their view, the Marxian analysis of capitalism demonstrates how the abusive exploitation of labour and land are two sides of the same coin. Historical materialism is thus profoundly concerned with ecology, yet refuses to reduce economic value to thermodynamics.

Finally, Marina Fischer-Kowalski has organized a group of scholars in Vienna working on 'social metabolism' through material flow analysis (Fischer-Kowalski and Haberl 1993; Fischer-Kowalski 1998; Fischer-Kowalski and Amann 2001; Eisenmenger and Giljum 2007; Weisz 2007). By translating international commodity flows into weight (i.e. tons), these researchers have been able to calculate the 'physical trade balances' of nations and even of the European Union. These calculations generally show that core nations import much more weight (materials) than they export, while the converse applies to nations in the periphery of the world system. Working on trade statistics from Colombia, Mario Pérez Rincón (2006) has shown how the country's deteriorating physical trade balance over several decades can be explained by the falling prices for its primary exports. He aptly uses the notion of an 'ecological Prebisch thesis' to indicate how structural inequalities in the world system have an ecological dimension that Prebisch (1950) and his colleagues did not fully recognize. Material flow analysis has also been used to show how core nations tend to export their most polluting industries to poorer nations (Muradian and Giljum 2007), illustrating how the import of high-quality resources and the export of degraded materials and health risks are two sides of the same coin.

My ambition has been to integrate all of these approaches into a single theoretical framework clarifying how societal relations of exchange and the material dimensions of production are intertwined in global metabolism (Hornborg 1992, 1998b, 2001b). As I show in Chapter 5, the study of unequal exchange of embodied labour can be combined with the study of unequal exchange of embodied land, demonstrating that the emergence of industrial capitalism in early nineteenth-century England was founded on a highly unequal exchange of both time (embodied labour) and space (embodied land) with its colonies. I refer to this basic condition of industrialized technology as 'time–space appropriation', suggesting also that David Harvey's (1996) observations on experiences of 'time–space compression' represent the phenomenological dimension of what in terms of political economy (and political ecology) can only be understood as a process of (time–space) appropriation. The rationale of industrial technology is to save time and space, but a global analysis reveals the extent to which this is achieved at the *expense* of (arguably 'underpaid'[1]) human time and natural space elsewhere in the world system.

To understand how industrial capitalism can present itself to us as 'efficiency' and 'progress', while we simultaneously acknowledge that it may be bringing us closer and closer to socio-ecological disaster, we need to grasp both the material, the social, and the cultural or cognitive dimensions of this mode of production. On the material side, we certainly need to be acquainted with Georgescu-Roegen's observations on entropy, Bunker's concept of an unequal exchange of energy, and Fischer-Kowalski's analyses of the uneven flows of matter in global society. To understand the structure of this global system, we cannot do without the world-system perspectives of Frank and Wallerstein. Finally, to grasp how it has been possible for us to be so deluded by our technology for over two hundred years, we need the Marxian analysis of fetishism. But although Marx was able to expose the fetishism of money and of commodities, he never fully recognized the fetishism of the machine. Yet it, too, is a material object attributed with magical powers that mystify unequal relations of exchange.

'Resilience' or collapse? The ideological disarmament of disaster

In very general terms, there seem to be three possible ways of relating to the issue of global socio-ecological sustainability. The first is to deny that there are any major problems that will not be solved by the market and new technologies (e.g. Lomborg 2001). The second is to acknowledge that current trends will lead to disaster unless major political or cultural reforms are achieved, but to continue to believe in the feasibility of such reforms (e.g. Gore 1993). The third is to acknowledge the risk of disaster, but as yet be unable to visualize any realistic strategy for preventing it. The first two attitudes are both optimistic regarding the future of global capitalist civilization, but for different reasons. To the extent that people in the industrialized North have thought about this issue, a majority today probably lean toward the second option. The fact that Lomborg's (2001) research has been officially rejected by a Danish research council, while Al Gore in 2007 was awarded the Nobel Peace Prize, seems to reflect widespread attitudes in core capitalist nations. Gore's movie *An Inconvenient*

Truth no doubt expresses the current world-view of most citizens of these nations. The problems cannot be denied, they seem to concede, but solutions are within reach. In fact, the confidence in solutions may often be a prerequisite for acknowledging the problems. Presentations of our global socio-ecological dilemma that do not offer feasible solutions will tend to be rejected, much like Karl Marx's analysis of capitalism probably would have been by the labour movement, if he had not simultaneously articulated a vision of socialism. This is a disturbing fact, since it makes intellectual sense to distinguish between the activity of describing a socio-ecological dilemma, on one hand, and offering practical or political solutions to it, on the other. Yet, option two is undoubtedly experienced by most people as morally superior to both option one and option three.

Against this background, it is not difficult to understand why the globalized discourse on 'sustainable development' within academia and management is dominated by assumptions of consensus and a trivialization of obstacles to implementation of the requisite policies. In order to remain within acceptable discursive territory, politicians and researchers alike are expected to assume a profoundly critical stance vis-à-vis current patterns of consumption, transports, and energy use, yet continue to offer pathways to sustainability that do not seem too uncomfortable or provocative. This explains why the rallying-cry of the early twenty-first century is not 'revolution' (as in the early twentieth century), but 'resilience'. The burgeoning discourse on 'resilience' in socio-ecological systems (e.g. Berkes and Folke 1998; Levin *et al.* 1998; Berkes *et al.* 2003) is an ideological phenomenon worthy of study in its own right. Intellectually, it is largely founded on the epistemological traditions of systems ecology and similar, natural-science approaches emphasizing the harmonious functioning of natural systems through adaptation, wise management, and appropriate technologies. Normative notions of strategies to promote the 'resilience' of socio-ecological systems exemplify this view. These approaches, and the agencies that continue to promote them, seem oblivious to several strong research traditions in the social sciences that have persuasively shown that socio-ecological systems are historically and currently characterized by structural problems of power, conflicts of interest, and unequal distribution.

Considerable amounts of economic resources are currently being invested in the idea that modern society will be able to learn how to manage natural resources sustainably by studying the small-scale, traditional ecological practices that for centuries have been systematically eclipsed by the logic of industrial capitalism (cf. Berkes and Folke 1998; Berkes 1999; Berkes *et al.* 2003). In the light of the alarming current trends of global environmental change (e.g. Jackson *et al.* 2001;

Millennium Ecosystem Assessment 2005), this notion seems naïve and paradoxical. Many anthropologists will feel strong ambivalence about such claims. Considering the fate of traditional or indigenous societies under extractive capitalism (cf. Nadasdy 2007:218), it is confounding to read that 'traditional resource management systems ... may have potential that has hardly been tapped' (Berkes and Folke 1998:13). Although there is definitely something to learn from local and contextualized ecological practice about the disastrous course of capitalist extraction, the suggestion that modern capitalist society might adopt the logic of its socio-ecological antithesis must be dismissed as mystification (Hornborg 1996). The only circumstance that might engender a revitalization of such local practices would be the collapse of capitalist civilization (cf. Tainter 1988), but we may rest assured that this is not the objective of all the research, conferences, publications, and policy statements that presently champion what Nadasdy (2007) calls the 'gospel of resilience'.

The explicit goal of the research on conditions of socio-ecological resilience is to incorporate such knowledge into 'resource management science' (Berkes and Folke 1998:1), yet the same sources explicitly *contrast* Western resource management science with 'traditional and local management' (ibid., 5). Here is another instance, then, of capitalism struggling to incorporate its antithesis. 'Resilience' is defined as the ability of a system to absorb disturbances before unpredictably changing its structure from one equilibrium state to another, less desirable one (ibid., 6, 12). The concept seems to have replaced the scientifically obsolete notion of the 'balance of nature' as the foundation of an ecological ethic (Nadasdy 2007:214). Although derived from the natural science of systems ecology (e.g. Odum 1989), the concept is applied to linked 'social-ecological systems' and thus obviously considered applicable also to social systems, much as 'classical human ecology' literature such as Park (1936) did not hesitate to apply ecological theory to societies (Berkes and Folke 1998:9). The proponents of resilience theory suggest that this dissolution of the boundary between nature and society is consistent with the way many traditional societies see their relationships with the environment, namely human society as part of nature rather than separate from it. This is a statement that should be approached with some caution, however. Although the dichotomy of nature and society indeed seems to be absent in many traditional cosmologies (cf. Descola and Pálsson 1996; Ingold 2000), these cosmologies generally have very little in common with systems ecology. To the extent that nature and society are conceptualized within the same framework, it is the social domain that is extended into the natural, rather than vice versa. Characteristic of these traditional cosmologies is not the existence of 'pre-scientific ecosystem concepts' (Berkes and Folke

1998:9), but the projection of social concepts of reciprocity, harmony, conflict, and power onto relationships between humans and non-human species (e.g. Descola 1994; cf. Chapter 7). To suggest that modern 'resource management science' emphasizing the resilience of socio-ecological systems is somehow cognate to pre-modern cosmologies, and sympathetic to the 'savage thought' of Lévi-Strauss (Berkes and Folke 1998:12–13), is thus a gross distortion. The claim to have rediscovered or reinvented indigenous ecological knowledge, legitimized as science, has been made repeatedly by resource administrators over more than a hundred years (Nadasdy 2007:211–212). The image of the ecologically conscious, pre-modern native has great symbolic power, but may here be resorted to as a way of erasing (or concealing) the historical heritage of resource management science as a means of restricting indigenous people's access to and use of the environment (ibid., 212–213).

Intimately connected with the notion of resilience is the idea of 'adaptive management' (Holling 1978). To the extent that adaptive management is defined by its capacity to handle 'surprises' or unpredictable events, it almost seems an oxymoron (cf. Nadasdy 2007:209–210). It is founded on the explicit belief that 'organizations and institutions can "learn" as individuals do' (Berkes and Folke 1998:10), but this is yet another highly problematic contention. It is one thing to observe that supra-individual knowledge systems can accumulate information in a manner analogous to individual learning, as is the case with both science and 'traditional ecological knowledge', but another to suggest that organizations and institutions (i.e. social structures) can learn. Social systems are not comparable to biological systems in this regard (Hornborg 2007:5). Their trajectories are generally propelled by individuals and groups struggling to maximize their power and affluence, yet there is no mention of power or contradiction in the 'analytical framework' for understanding social-ecological systems (Berkes and Folke 1998:15). In view of the recurrent catastrophes that have historically afflicted human societies at all levels of integration, from bands to empires, it is very difficult to subscribe to this notion of societal learning. To the extent that social systems are able to draw on past experience to avoid future disaster, it is important to recognize that such 'learning' is a matter of storing information as practically useful knowledge, rather than of mystically codifying 'wisdom' into the cybernetic structure of social organization. The proponents of resilience theory are unclear about which of these two concepts of societal learning they are advocating.

Several social scientists have identified such weaknesses in resilience theory (Hanley 1998:248; Lélé 1998:253; Nadasdy 2007). As Lélé (1998:253) remarks, 'The lessons from the history of environmental policy are unambiguous: no amount of "trust", "clever institutional design"..., or "epistemic consensus" ... can compensate for major asymmetries in the interests and powers of the different actors'. Nadasdy (2007:216) observes that although proponents of adaptive management

> clearly recognize that it is the economic imperatives of modern extractive and agro-industries that are the root cause of the management 'pathologies' that lead to decreased resilience and ultimate collapse...,
> their proposed solutions do not address these larger issues at all.

Indeed, he continues, it is largely because of the exploitative dynamic of capitalism that conventional resource management even exists, and the focus on resilience 'has the implicit goal of maintaining the social-ecological relations of capitalist resource extraction and agro-industry' (ibid., 217–218).

The discourse on resilience is conspicuously oblivious not only of power, conflict, and contradiction, but also of culture. The institutions and knowledge systems that Berkes and Folke (1998:6) refer to as 'cultural capital' are acknowledged as significant components in socio-ecological systems only inasmuch as they contribute to resilience, i.e. in the obsolete and functionalist sense in which culture was conceived in mid-twentieth-century schools of anthropology such as cultural ecology and cultural materialism. There is no discussion of the role of cultural idiosyncrasies (e.g. various versions of fetishism and consumption patterns) as autonomous driving forces propelling the ecological trajectory of a social system. 'Culture' in the modern anthropological sense is obviously not acknowledged as a relevant component of socio-ecological systems. Yet, as we have seen, culture is in many ways crucial to socio-ecological processes and relations. It shapes our way of thinking and talking about sustainability and exchange, including our categories of 'technology', 'economy', and 'ecology'. It generates our specific kinds of fetishism and consumption, determining which aspects of material culture that we consider indispensable. It prompts us, in different contexts, to attribute productivity or 'value' to different factors of production. In fact, it is finally also responsible for our modern fascination with 'traditional ecological knowledge', which paradoxically has become so central to the 'gospel of resilience'.

I do not doubt that the proponents of 'resilience theory' are sincerely concerned about the destructive course of industrial capitalism, and I naturally agree with them that most non-modern systems of resource management perform better in terms of sustainable human–environmental relations, but in order to seriously scrutinize the socio-ecological logic of the modern world the discourse on resilience cannot ignore the many relevant insights on this logic that derive from social and cultural theory. Should it continue to do so, it will remain as incapable of transforming capitalist practice as the old romantic stereotype of the 'ecological native' to which it is an heir, and it will not pose the slightest challenge to business as usual.

The ideology of ecological modernization

In this chapter, I have discussed various ways in which conventional discourse on sustainability fails to acknowledge the distributive, political, and cultural dimensions of global environmental problems. Central to the argument is the phenomenon of 'machine fetishism', i.e. the inclination to view the technological capacity of a given population as independent of that population's position in a global system of resource flows. Machine fetishism is an ideological illusion maintained by keeping perspectives from the natural sciences, social sciences, and humanities effectively separated, and by perceiving the operation of global market price mechanisms as tantamount to guaranteeing reciprocal exchange. The three main ingredients in the theory of machine fetishism – the acknowledgement of ecologically unequal exchange, world-system analysis, and the concept of fetishism applied to political economy – can all be traced back to Karl Marx. The 'zero-sum game' view of technological and economic growth, which underlies all these three ingredients, had been consolidated in Marx's concepts of capital accumulation and the metabolic rift already in the mid-nineteenth century, as Europe was entering the age of fossil fuels. As I will elaborate in Chapter 6, the current concerns about peak oil and climate change suggest that we may be in the midst of a historical transition (back) to an economy based primarily on productive land area. This prompts us to reconceptualize the unequal exchange of embodied labour and embodied land in terms of 'time–space appropriation', and to question general theories of value attempting to account for economic growth with reference to a specific factor of production, whether labour, land, or capital. Such theories (e.g. labour theories of value, Physiocracy, modernization theory, etc.) should thus be viewed as geographically and historically positioned social constructs. The illusion of machine fetishism generates a representation of socio-economic differences in geographical space as if they were differences in historical time, epitomized in mainstream conceptions of 'development'. The discourse on 'sustainable development' is thus confronted with the political challenge of representing solutions to glaring problems of global inequalities in terms of consensus, benevolent 'management', and the restoration of socio-ecological 'resilience'. Although the bulk of the insights that we need to understand the current political ecology of the world system can be derived from the Marxian framework, we are today in a better position to discern the pervasive logic of environmental load displacement and thus also to deconstruct the ideological faith in technological solutions.

The various epistemological challenges presented in this chapter can thus be viewed as interconnected illusions postponing systemic crisis and obstructing rational societal negotiations that acknowledge the political dimensions of global ecology. These challenges are nothing less than the ideological pillars of ecological modernization. They are five in number:

1 The fragmentation of scientific perspectives into bounded categories such as 'technology', 'economy', and 'ecology'.
2 The assumption that the operation of market prices is tantamount to reciprocity.
3 The illusion of machine fetishism, i.e. that the technological capacity of a given population is independent of that population's position in a global system of resource flows.
4 The representation of inequalities in societal space as developmental stages in historical time.
5 The conviction that 'sustainable development' can be achieved through consensus.

These five pillars of ecological modernization are serially interconnected in the following manner: 1 and 2 are prerequisite to 3; 3 is prerequisite to 4; and 2 and 4 are prerequisite to 5. The aim of this chapter has been to indicate some ways in which these illusions can be unravelled.

Beyond these illusions, what might truly sustainable socio-ecological relations look like? Although hardly comfortable – and likely to be dismissed by most readers as unfeasible – my conclusion is that the only way of achieving 'sustainability' would be by transforming the very idea and institution of money itself (see Chapter 8). The fundamental problem is that our global social system has adopted an institution, based on the idea that *everything is interchangeable*, that continues to encourage an accelerating dissipation of resources. General-purpose money inexorably rewards such dissipation by providing access to more resources to dissipate. Only by breaking that logic will human society be able to gain mastery over its own mastery over nature.

2 Fetishism, dissociation, and the cultural analysis of capitalism

In the previous chapter, I showed how the concept of fetishism can help explain why most modern people tend to think of new technological inventions as signs of human progress, rather than implying accumulation and unequal exchange. In this chapter, I will discuss more extensively how the way humans relate to material objects is a fundamental aspect of industrial capitalism. The aim is to reconnect the discourse on 'fetishism', the main thrust of which has become largely restricted to exploring the personal phenomenologies of aesthetic experience (cf. Apter and Pietz 1993; Spyer 1998; Mitchell 2005; Latour 2010),[1] to a general critique of global capitalist relations. For obvious reasons, the ambition here is not to attempt to review the voluminous discourses on fetishism, animism, epistemology, magic, materiality, technology, or consumption, but to bring together insights from these various topics to suggest new ways of illuminating some cultural dimensions of modernity and capitalism. More specifically, the goal is to combine some relevant aspects of culture theory with perspectives from political economy, world-system analysis, and ecological economics in order to 'defamiliarize' (Marcus and Fischer 1986) our everyday understanding of technology. Empirically, the discussion ranges from early British textile factories and the Luddite movement to indigenous Amazonian animism and ancient Andean ritual.

Expanding the Marxian concept of fetishism

Karl Marx famously observed that relations between people in capitalist society assume the form of relations between things. The fetishism of money and commodities thus obscures the social foundation of these objects, as a result of the alienating split between people and the products of their labour. It simultaneously animates such things, by attributing to them autonomous value, productivity, or growth. To deconstruct fetishized human–object relations such as these, in order to reveal underlying social asymmetries, can be a powerfully subversive analytical strategy. It helps us to understand phenomena as diverse as the pervasive desire for consumption goods and the violence of physical sabotage. Ultimately, it provides a radically alternative perspective on the economic, political, and environmental inequalities of global society. In order to seriously challenge those inequalities, we would have to open our eyes to the social relations underlying technology. As I argued in Chapter 1, machines[2] are basically also inanimate objects attributed with autonomous productivity, obscuring their own foundation in asymmetric global relations of exchange. The political innocence of 'technological progress' qualifies as a 'world view' in Kearney's (1984) sense. As it is fundamental even to a Marxian perspective, it poses a peculiar contradiction to social science drawing on Marx's analyses of capital: How can capital, once it assumes the form of technology, become exempt from political critique?

As we shall see in this chapter, recognizing human–object relations as mystified social relations can actually also help us reconceptualize our insatiable appetite for commodities as a subtle transformation of cannibalism. In order to understand fetishism as a simultaneously cultural and political phenomenon ultimately implicating macro-scale power structures at the global level, we need to consider a spectrum of ways in which humans can relate to other beings and other things in their surroundings. Let us begin by discussing the contrast between animism and objectivism.

Animism versus objectivism: modernity as dissociation

The topic of animism continues to intrigue modern people. What, then, do we mean by 'modern'? As a number of social theorists have suggested, the social condition and technological accomplishments of European modernity have been founded on a categorical (so-called Cartesian) distinction between Nature and Society. It is by drawing a boundary between the world of objects and the world of meanings that the modern project has emerged. By, as it were, distilling Nature into its material properties alone, uncontaminated by symbolic meanings or social relations, modernists have been freed to manipulate it in ways unthinkable in pre-modern contexts. Objectivism thus suggests a kind of moral or emotional *dissociation* from that part of reality classified as object.

Animism suggests the very antithesis of this objectifying modern stance. Yet it is not a phenomenon that can be relegated to a previous period in human history. As Descola (1994), Bird-David (1999), Ingold (2000), and other anthropologists have shown, many contemporary people who are intimately engaged in gaining their subsistence from local ecosystems continue to approach their non-human environments through what is now being called a 'relational' stance. Entities such as plants or even rocks may be approached as communicative subjects rather than the inert objects perceived by modernists.

We can approach the notion of animism from the perspective of what Latour (1993) has called a 'symmetric anthropology': an anthropology that does not merely represent an urban, modern perspective on the premoderns in the margins, but that is equally capable of subjecting modern life itself to cultural analysis. Latour argues that we have, in fact, never been modern. The notion that the world of objects and the world of subjects are separable, in any other than an analytical sense, has been an illusion from the start. Not only do human beings everywhere impute personhood and agency to entities that according to official modernist doctrine ought to be classified as objects (think of our favourite trees, houses, cars, or teddy bears), but Latour makes the important point that modernity itself, through the new socio-technical networks unleashed by its dualist epistemology, continually generates ever more obvious examples of hybrids or 'quasi-objects' that contain both subjective and objective aspects, and that span the divide between Culture and Nature. In other words, the official Cartesian ideology of subject–object dualism is not only contradicted in our everyday lives; when applied in social and technological practice, it inexorably produces increasingly conspicuous evidence of its own invalidity. From the ozone hole to genetically modified organisms, the real world afflicted by modernity has shown itself to be not only permeable to, but imbued with, politics, meanings, and human intentions. The new technologies and networks prove to be not objects but what Latour calls quasi-objects: part Nature, part Society, and brimming with agency.[3]

Our training in the skills of modernist detachment and objectification is contextual, as illustrated by the professional logger who privately cares for his garden, or the industrial butcher who privately cares for his dog. The efficacy of modernity in unleashing wholesale transformations in human–environmental relations lies in the creation of a spectrum of highly specialized occupations, each emphasizing its own specific application of objectification and detachment, so that the total impact of modern society is unrestricted by moral concerns, while each individual is able to maintain, by and large, a moral identity. In other words, it may not so much be an incapacity to relate as such that distinguishes us from the animists, as the incapacity to exercise such relatedness within the discursive and technical constraints of the professional subcultures that organize the most significant share of our social agency. Science and technology do not so much make us into robots, as make specific parts of our behaviour robot-like.

Or does the difference indeed go deeper than this? Do modern people have a generalized tendency to perceive their environments as collections of objects? A compelling observation in this direction is Viveiros de Castro's (1999) intriguing suggestion that Europeans and Amerindians have diametrically opposite images of how humans and animals are constituted. Whereas Europeans tend to conceive of human beings as biological organisms masquerading in a cultural costume, Viveiros de Castro observes that Amazonian Indians view animals as fundamentally persons concealed under their animal surface. For Europeans, then – or at least for European biologists – living things are fundamentally objects, while for Amazonian Indians they are fundamentally subjects. The latter, of course, is an unusually concise statement of animism. But then, the Cartesian view of non-human animals seems counterintuitive, even to the most ingrained modernist. A telling illustration of this is the classical example of the early European vivisectionists, who felt compelled to sever the vocal cords of the dogs whose living anatomy they explored (cf. Evernden 1985:16–17). In effect, they could only perform their modernist task after having shut off the communicative link – the relation – between dog and human. This is a very tangible illustration of how objectification can be seen as a strategy of dissociation.

Modern struggles for relatedness: the semiotics of knowing

Would modernity be impossible in a world where living things are consistently recognized as subjects? Latour's answer seems to be yes. It is only by severing or submerging our capacity for 'relatedness' that we are set free to impose our modernist designs on the world. Significantly, to make this point he refers to Descola's (1994) suggestion that traditional societies of Amazonia retain their relative inertia – compared to Europe – precisely because their conception of the non-human environment remains embedded in their moral conception of society (Latour 1993:42). Animism, to Descola, is the projection of social metaphors onto relations with the non-human world. In not separating Nature and Society, Amazonian Indians like the Achuar automatically embed their ecological practice in a compelling moral system. For centuries, mainstream European society has refused to be thus constrained, and this liberation of capitalist modernity has been founded on the incommensurable distinction between Nature and Culture. Against this background, it seems ironic that calls are now being made for an 'environmental ethics'. The very idea indeed poses a conundrum for Cartesian objectivism. How shall we be able to reintroduce morality into our dealings with our non-human environment, now that we have invested centuries of training and discourse into convincing ourselves that Nature lays beyond the reach of moral concerns?[4]

If the systematic modern denial of relatedness is somehow at the root of ecological crisis, as many environmentalists believe, what are our prospects for resurrecting it? It is difficult to imagine that modern society as an act of instrumental reason should begin inculcating in its citizens the long-term ecological validity of pre-modern metaphors of what Bird-David (1993) calls 'subject-subject relatedness'. However much we admire the eco-cosmologies of the Nayaka, the Achuar, or the Cree, we should not expect to encounter them anywhere but in the anthropology departments, and definitely not in mainstream textbooks in ecology or sustainability.

Nor can we put much hope in what has become known as New Age spirituality. The movement as such is highly relevant from the point of view of this discussion, but it is a flimsy platform on which to build a future: a post-modern symptom of epistemological collapse rather than an advance on modernity. When neo-pagans and other New Age enthusiasts proclaim that this or that sacred site possesses such strong 'energy', it seems as if they are indeed struggling for relatedness – for a restoration of meaning beyond the existential wasteland of modernity (cf. Heelas 2008) – but remain confined to the modernist (and, in fact, scientistic) vocabulary through which objective properties are attributed to distinct, external things (cf. Hanegraaff 1998; Hammer 2000). New Age spirituality thus reproduces the inclination of mainstream science to turn subjects into objects, but now the objectified subject(ivity) is one's own.

Surrounded by philosophers and sociologists of science announcing the end of Cartesian objectivism and acknowledging the extent to which human meanings infuse the material world, anthropologists discussing animistic understandings of nature will now be excused for taking them more seriously than a generation ago. But rather than going native, or adopting some version of New Age spirituality, it is incumbent on us to analytically sort out what epistemological options there are, and to ask why pre-modern, modern, and post-modern people will tend to deal with subject–object relations in such different ways.

The object – in the sense of a material, intrinsically meaningless, but essentially knowable reality – appears to be a thoroughly modern invention. Whichever interpretative schemes conventionally adhered to in pre-modern societies, they enjoyed a kind of immutable authority that modern knowledge rarely achieves. It is the predicament of modern people to remain chronically uncertain about the validity of their own representations. This modern condition of reflexive uncertainty can either be harnessed in the production of new but provisional certainties (as in science) or assume the form of solipsism, disengagement, and indifference. The latter alternative is what we have come to know as the post-modern. It is a condition where the exhausting attitude of chronic scepticism tends to give way to a kind of resigned gullibility. All hope of certainty has vanished, but precisely because no pretence to power or truth can be admitted, any pretence is as good as any other (e.g. the claims of the neo-pagans). As in the pre-modern condition, a sign is again naively perceived as an unproblematic index of identity – rather than an arbitrary symbolic convention demanding to be challenged – but now simply by virtue of positing itself as such, rather than because of an assumed correspondence with some underlying essence (Hornborg 1999a). This postmodern abandonment of essence is what Baudrillard (1975) has aptly called the 'autonomization of the signifier'.

The problem with objectivism – as unimaginable for the pre-moderns as it is unacceptable for the postmoderns – is the notion of a kind of knowledge that is not situated as part of a relation. By posing as disinterested representation, decontextualized from any political aspirations, modernist knowledge production suggests a relinquishment of responsibility, but in fact serves – through technology – to set the instrumental rationality of the powerful free to go about its business in the world. But the postmodern mirror-image of objectivism – that is, relativism – certainly fares no better in terms of responsibility. Both these epistemologies have been spawned by the same, modern subject–object dichotomy. The division into natural versus human sciences, pitting realism against constructivism in Western knowledge production, remains a projection of this fundamentally existential, dualist scheme. The former takes the represented object as its point of departure, the latter the constructing subject, but neither acknowledges their recursivity, i.e. their relation. One reason why animism continues to intrigue us may be that this is precisely what animism does. Rather than viewing knowledge as either representation or construction, animism suggests the intermediate view that knowledge is a relation that shapes both the knower and the known. An animistic or relational ontology is a mode of knowing that is not only constitutive of both the knower and the known – as is all knowledge, according to the cognitive scientists (cf. Maturana and Varela 1992) – but that crucially also acknowledges this fundamental condition, and thus also the responsibilities that must always adhere to the very act of knowing. Beyond objectivism and relativism, there can only be relationism. Perhaps because purely instrumental knowledge and rational risk assessment can never be as powerful incentives for human action as moral imperatives, we may need new metaphors capable of sustainably relating us to the rest of the biosphere (Hornborg 1996).

Animism raises our curiosity as the hesitant acknowledgement of suppressed childhood experiences, the assertion of which would challenge the entire modern project. Relatedness is a condition that most of us continue to be capable of achieving in particular, experiential contexts of some minimal duration. Our modernity – our inclination toward abstraction, detachment, and objectification – is the product of our disembedding biographies (Hornborg 1999b). It is in being involuntarily deprived of relatedness that we become Cartesianists. The powerful historical trajectory of objectivism relies on a peculiar recursivity between social disembeddedness, Cartesian epistemology, and technology – ultimately, that is, between individual existence and socio-technical power structures. The epistemological predicament articulated by Descartes was not so much an innovative, cognitive shift from animism to objectivism, as the emergence – or unprecedented generalization – of a social condition of alienation. Through processes of increasing commodification and alienation – and more recently through the proliferation of new technological hybrids in Latour's sense – the social condition of modernity has accentuated our anxieties about where or how to draw boundaries between persons and things (cf. Kopytoff 1986), amplifying a pervasive (Cartesian) dissociation of self from non-self that, as we have seen, is at the root of both solipsism and objectivism. To the extent that we do continue to animate our favourite trees, houses, cars, or teddy bears, it is because we continue to need concrete reference-points onto which to anchor our selves (cf. Shweder and Bourne 1984; Hornborg 2001b:206–208). It is the long immersion in the concrete and experiential specifics of place that yields conditions conducive to relatedness – vis-à-vis irreplaceable persons, localities, and things. This, if anything, should provide us with clues about the prospects for resurrecting relational ontologies.

Fetishism and the cultural analysis of capitalism

Let us now turn to a second and complementary aspect of our argument, with the aim of bringing both parts together further on. There is another and thoroughly modernist way in which things can be animated, which has to do not with experiential resonance but with ideology and political economy. Animation is in fact fundamental to fetishism, and fetishism to Karl Marx was central to modern capitalism (cf. Friedman 1974; Taussig 1980; Pietz 1985–1988, 1993, 1998; Miller 1987, 2005; Ellen 1988; Hornborg 1992, 2001a). It is indeed important to ask how animism relates to fetishism. There is a crucial difference between representing relations between people as if they were relations between things (Marxian fetishism), and experiencing relations to things as if they were relations to people (animism). The former is an ideological illusion underpinning capitalist political economy, the latter a condition of phenomenological resonance. We should probably further distinguish between the animation of living things such as trees (animism, more narrowly defined) and that of non-living things such as stones or machines (that is, fetishism). Cartesian objectivism and fetishism here emerge as structural inversions of one another: the former denies agency and subjectivity even in living beings, whereas the latter attributes such qualities to inert objects. In this framework, a more strictly defined category of animism would be reserved for the intermediate and quite reasonable assumption that all living things are subjects, i.e. equipped with a certain capacity for perception, communication, and agency. Animism, fetishism, and objectivism can thus be understood as alternative responses to universal human problems of drawing boundaries between persons and things.

Perhaps some of these problems can be alleviated by recognizing the difference between drawing boundaries in an analytical and an ontological sense. We can probably all agree with Latour that Nature is continually being intertwined with Culture or Society in our landscapes, our bodies, and our new hybrid technologies that obviously invalidate ontological versions of the Cartesian dichotomy. But does this mean that the categories of Nature and Culture, or Nature and Society, are obsolete and should be discarded (cf. Ingold 2000)? On the contrary, never has it been more imperative to maintain an *analytical* distinction between the symbolic and the pre-symbolic, while acknowledging their complex interfusion in the real world. Only by keeping Society and Nature analytically apart can we hope to progress in the demystification of that hybrid web in which we are all suspended, and which more than anything else obstructs our pursuit of relatedness: the realm of animated objects that we call *technology*. We more than ever need to retain our capacity to distinguish between those aspects of technology that derive from Nature and those aspects that derive from Society. The Laws of Thermodynamics and the political economy of oil prices require completely different analytical tools.

What is sometimes referred to as the 'anthropology of technology' comprises some interesting attempts to explore the interface of culture and materiality, but the field tends to be conspicuously detached from considerations of global political economy (cf. Latour 1996; Ingold 2000; Ihde and Selinger 2003).[5] For instance, both Ingold and Latour are preoccupied with the dubious modern distinction between persons and objects and between Culture and Nature, both recognize that this distinction is paradoxically itself cultural (cf. Latour 1993:99; Ingold 2000:42), and both keep returning to the phenomenon of modern technology as an arena where the distinction becomes blurred or at least problematic, but neither of them is concerned with how this very arena is itself a manifestation of global rates of exchange.

There is something about the general concept of 'technology' that seems to escape us, both as social scientists and as citizens. On one hand, modern technology seems quite obviously to be a strategy for capacitating an affluent minority of the world's population through an asymmetrical exchange – an expanding net appropriation – of resources from the rest of the world (Hornborg 1992, 2001b). On the other hand, technology tends to be represented as a politically innocent and intrinsically productive union of human inventiveness and the pure material essence of Nature – indeed as a gift of the wealthier, developed nations to the rest of humanity (Adas 1989, 2006; Adams 1996; Marsden and Smith 2005; Friedel 2007; Headrick 2010). How are these two contradictory images of technology able to coexist, without the former contaminating the latter? The answer, Latour would undoubtedly say, lies again in that rigid categorical distinction between Nature and Society, between the world of pure objects and the world of human relations. Once classified as object, technology is automatically immune to political critique. For how could pure objects be conceived as sources of malign agency? If the behaviour of the early nineteenth century Luddites today strikes us as odd, it is because they were not yet quite modern. Today we supposedly know better than to direct our political frustrations at machines. The efficacy of technology, we hold, comes from 'objective properties intrinsic to the nature of things' (cf. Latour 1993:51). Like economic rationality and scientific truth, says Latour, technological efficiency 'forever escapes the tyranny of social interest' (ibid., 131).

But if these modernist convictions were indeed to collapse, and we should realize the extent to which our technologies are in fact politically constituted, our machines would cease to be pure objects and conceivably be accredited with a malicious agency far surpassing that of any premodern fetishes.[6] Such a transformation would hinge on the nature of the agency attributed to technological objects: from having been fetishized into politically neutral, autonomous agents, they would emerge as social manipulations. To expose the agency of these cornucopian 'productive forces' as a transmutation and deflection of the agency of other humans would be to render morally suspect that which modernity had couched in the deceptive neutrality of the merely technical. And in seeing, for the first time, the machines as they really are – as machinations – perhaps the animist within us would stir again, and we would ask ourselves: What manner of creatures *are* these things, part mineral, part mind, that serve the few to enslave the many, while fouling the land, the water, and the air?

Such a scenario serves to remind us that animism and relatedness bring possibilities not only of harmony and community, but also of horror and rage. It might help us understand how the Cartesian suppression of relatedness has served a fundamental ideological purpose in the emergence and expansion of industrial capitalism. Against this background, nothing would appear to be more revolutionary than to try to rekindle some of our pre-modern attitudes as we confront the demons of our own making.

Fighting against machines

History tells us, however, that revolutionary rage is rarely enough. Machines became the objects of political violence exactly two hundred years ago, in early industrial England, through the so-called Luddite movement (Sale 1995; Fox 2002; Binfield 2004; Jones 2006). This short-lived movement created considerable turbulence in the heartland of early British industrialization (the counties of Yorkshire, Lancashire, Cheshire, Derbyshire, and Nottinghamshire) from late 1811 to early 1813. Thousands of local, proto-industrial textile workers who had seen their livelihoods eclipsed by the large-scale machinery of factories perceived these new buildings and their technologies as immoral contraptions violating traditional principles of justice and fairness. The factory system was explicitly likened to 'colonial slavery' (Sale 1995:23), and the embittered workers who suffered dwindling incomes and unemployment responded with revolutionary fury. Their response, which may then have appeared somewhat less futile than it does today, was to attack and destroy the machines themselves. In slightly over a year, damages to technological infrastructure exceeded £100,000, and many factory owners were attacked and injured. As Charlotte Brontë later wrote, it is not difficult to understand why 'these sufferers hated the machines which they believed took their bread from them; they hated the buildings which contained those machines; they hated the manufacturers who owned those buildings', yet 'it would not do to stop the progress of invention' (quoted in Sale 1995:15–16). The British authorities swiftly crushed the Luddite movement and executed its leaders. Sale (1995:5) concludes that 'the architects and beneficiaries of the new industrialism knew that it was imperative to subdue that challenge, to try to deny and expunge its premises of ancient rights and traditional mores, if the labour force were to be made sufficiently malleable, and the new terms of employment sufficiently fixed, to allow what we now call the Industrial Revolution to triumph unimpeded.'

Did the Luddites in 1812 really perceive sabotage of machinery as a possible way of intervening in the logic of capitalism? Or were they driven by the same kind of rage that has repeatedly prompted embittered people to destroy the fetishized monuments and images (the ritual 'technologies') of pre-modern elites such as those of ancient Rome, the tenth-century Maya, or Easter Island?[7] Are the two incentives in fact inseparable?

Whatever the case, historical hindsight suggests that theirs was simply not a feasible strategy. Sale (1995) aptly titled his book *Rebels Against the Future: The Luddites and their War on the Industrial Revolution.* The social project of physically destroying machines, however, deserves reflection. In directing their anger at these mechanical objects, the Luddites can be said to have engaged in a pre-modern form of fetishism, if we define fetishism as the attribution of animacy and agency to non-living things. On the other hand, their actions can be said to have been intuitively justified, if we concede that the machines were indeed objectified social relations. In fact, to ontologically divorce the machines from social relations of exchange, as has become the predominant outlook in modern industrial society, suggests a more subtle and mystifying form of fetishism (Hornborg 2001b). If the (unevenly distributed) accumulation of technological infrastructure is viewed simply as what Brontë called 'the progress of invention', i.e. as an inevitable accretion of ever greater technological capacity, irrespective of a population's position in a global system of resource flows, we may be mistaking a privileged position in global social space for an advanced stage in historical time (cf. Fabian 1983). Nevertheless, the persuasiveness of machine fetishism has proven highly durable for two hundred years. The 2001 attacks on the World Trade Center in New York notwithstanding, physical sabotage against modern infrastructure has not been a viable strategy for subverting modern power.

The concept of fetishism as used by Karl Marx helps us to see how human' relations to objects are ultimately about their relations to other humans. From this point of departure, David Graeber (2001, 2007) has used ethnographic material from more than a century of comparative anthropology to subvert mainstream modern conceptions of economy and power. In pursuing such a strategy, he is actually following in the footsteps of Marx himself, who turned the notion of fetishism back on Europeans attributing it to exotic Others. It seems that relations of social power are more or less universally mediated by fetishized objects (cf. Friedman 1974; Taussig 1980; Godelier 1986; Bloch 1989; Latour 2010). The human exchange and appropriation of things has always been about the production of persons, but, as Graeber reminds us, commodity fetishism encourages us to imagine otherwise. For instance, although the idea of private property is a thoroughly social relation, i.e. a person's right to exclude others from access to a thing, it presents itself to us as a relation between that person and that thing. Nor do we generally see that the commodity is an embodiment of other people's labour and landscapes. If the consumer's sovereignty over his or her commoditized objects is modelled on the monarch's sovereignty over his or her subjects, as Graeber (2007:73) suggests, the affinity between the two relations thus boils down to a transformation of social power.

Viewed in this light, it is revealing to see how closely related capitalism is to slavery (Graeber 2007:85–112). Not only was wage labour generally abhorred by free men in pre-capitalist societies, the underlying principles of capitalism and slavery are disturbingly similar: both systems involve exchanging abstract human labour power for money; both involve rationalized industrial organization of production; both rely on a spatial separation of the reproduction of labour and its realization in production; and both (in principle) deny any relevance of a worker's social identity outside the workplace. (Yet, Graeber notes, capitalism presents itself to us as freedom: the freedom of every individual to *sell* his or her freedom.)

What do all these features shared by capitalism and slavery have in common? The commoditization of abstract human labour power achieved something familiarly modern by dissociating (Karl Polanyi or Anthony Giddens would have said 'disembedding') productive activity from all other aspects of human life. In thus systematically alienating human beings from the products of their labour, these systems make it possible for the extracted agency or life force (Graeber calls it 'creative action') of human workers to be appropriated by others in the form of objects representing congealed abstract labour. This is the foundation of Marx's concept of commodity fetishism and in itself a powerful step toward a cultural deconstruction of the naturalness of capitalism. The abstraction of labour power is a means of transforming human energy into profits, or capital.

But the analysis can be pushed even further, for paradoxically, Marx was not a Luddite.[8] The very tangible, material operation of what we think of as modern 'technology' is no less than commodities an embodiment of the deflected agency or life force of human workers. As we suggested in the previous chapter, every 'technological' solution is ultimately a social relation in the sense that it will have implications for the societal distribution of the burden of problem-solving. The car or computer that saves its owner time represents losses of time for the myriad workers (such as in mines or oil fields) whose congealed labour it represents. Moreover, to the extent that modern technologies make possible a more efficient use of urban or agricultural space, for those segments of global society who can afford them, it is important to consider that they may represent losses of natural space (such as for strip mines or oil fields) elsewhere on the planet.

It is not a coincidence that Harvey (1996) and Graeber (2001) have both been intrigued by Munn's (1986) study of how the exchange of premodern money objects in Melanesia generates specific kinds of 'space-time' in the sense of particular relations of claims and reciprocities that variously extend the social reach of individual persons. The transition to general-purpose money, which now dominates our lives, has entailed a generalization of exchange in time and space. Money can thus be seen as an institution for delaying and extending the universal, gravitational pull of reciprocity. It is the transformations of this pull that make possible increasingly asymmetric exchanges between distant populations and ecosystems. And it is these increasingly asymmetric exchanges that make possible our technology. Time–space compression relies on time–space appropriation.

Machines, magic, and power

In struggling to grasp and communicate the magnitude of our misunderstanding of the phenomenon of technology, I have found no better framework than the Marxian concept of fetishism (Hornborg 1992, 2001b). As also Graeber (2001:239–246) has recognized, this concept immediately prompts us to consider the issue of magic.[9] Magic and power share the same hybrid position between scam and efficacy. There is an important sense in which the seemingly inexorable, material logic of the capitalist world order ultimately rests on the beliefs and conceptions of its participants. If restricted to the psychology of financial collapse, this is quite obviously true. But I would seriously argue that even technological efficacy is cognate to magic. I have repeatedly found that the quickest way to communicate what I mean by the concept of 'machine fetishism' is to compare modern technology with the magical 'soup stone' of European folklore:

> A hungry tramp is reluctantly admitted into a rural kitchen, but the housewife has no intention of serving him any food. He pulls a stone out of his pocket, asking merely for a pot of water to boil some soup on it. The housewife is too intrigued to deny his request. After a while, stirring and carefully tasting the water, the tramp observes that the soup might be improved with some flour, as if this was the only missing ingredient. The housewife, still baffled, consents to offer him some. Then, one by one, he similarly manages to lure her to add the various other ingredients, until finally she is amazed to find a delicious soup *cooked on a stone*.

In transferring attention from the wider context to its imaginary centre, the stone in the soup is the prototypical fetish. Fetishized objects are in an important sense constitutive – not just misrepresentations – of accumulation and power. They are visualized as intrinsically generative or productive, and they are indeed responsible for processes of accumulation, but only by orchestrating them, whereas this orchestration itself hinges precisely on obscuring their social basis in unequal exchange. No more than the stone contributed to the soup is a fetishized sacred king like the Inca emperor the source of his people's affluence (Godelier 1986; Hornborg 2000). Similarly, the industrial machine (i.e. the technological object) is but a fetishized node in a global system of resource flows. If those flows were to cease, the machine would grind to a halt.

Now that I have said that modern technology works like magic, I suppose it is incumbent on me to clarify the difference between them. What is the difference between the efficacy of magic and the efficacy of the machine? This question is simultaneously the question of how pre-modern sacred and ritual power could be transformed into modern economic and technological power. Graeber (2001) shows that the former never operates only by power of suggestion, or, conversely, by power intrinsic to the people who have it, but through the social relation by means of which these people are able to convince others that they do have it. Pre-modern power is thus cognate to magic because both rest on the premise that stage illusions work and must thus on some level be true. The import of fetishes such as *Spondylus* shells to ancient Cuzco helped the Inca court to convince the emperor's ten million subjects that his ritual communication with his father *Inti* (the Sun) was the prerequisite of agricultural productivity, and that it was entirely appropriate for them to reciprocate by spending significant amounts of their time working his fields and building his terraces (Murra 1962, 1975; Paulsen 1974, 1977; Marcos 1977/1978; Morris 1978; Godelier 1986). The illusion no doubt *worked* in the sense that the metabolism of the Inca empire hinged on the flow of red oyster shells from Ecuador to Cuzco. It was ultimately dependent on the trade of pre-modern prestige goods along the west coast of South America, through which merchants were able to acquire *Spondylus* in exchange for other objects coveted in Ecuador, such as packs of axe-shaped pieces of copper (Shimada 1985, 1987). This trade, of course, hinged on the cultural evaluations that determined the rates at which copper could be exchanged for *Spondylus* (Rostworowski 1977; Salomon 1986; Hornborg 2000).

Industrial technology, no less than theocratic ritual, is dependent on such pivotal exchange rates (e.g. oil prices). The difference is that, in industrialism, the transformation of imports into work has been locally objectified (into technology) so as to seem entirely 'material' and non-social. But what has actually happened is that the pivotal 'evaluative' moment has been shifted from the local to the global level. Locally, it has been delegated to the non-negotiable, kinematic logic of machines, but these are in themselves manifestations of global exchange rates. In ancient Peru, what were imported over longer distances were primarily symbols, which were ritually convertible into work in the form of manual labour. The productive potential on which the system was based was still local labour, and the rate at which prestige goods were converted into work was to some extent negotiable. But in modern industrial centres, it is increasingly the productive potential itself that is imported, which means that the imports are *physically* convertible into work, and that it is the global rather than the local conversion rates that ultimately determine the feasibility of accumulation. No less than ritual, machines mystify us by pretending to be productive independently of exchange rates. In modern capitalism, however, the mystified exchanges have become even more opaque, and the magic agency of fetishized objects has become compelling in completely new ways.

Modern power based on economic and technological accumulation is thus, like pre-modern power, dependent on the ability of social elites to extract obedience and labour energy from the myriad human beings who provide them with the means of asserting these demands, and for this reason remain as dependent on monopolies on legitimate coercion. They continue to operate only as long as the people they control can be persuaded, by magic and/or coercion, to subscribe to the claims to power offered by the elite. At this moment in history, these claims hinge, for instance, on the promises of continued economic and technological growth, and of global sustainable development. Perhaps most centrally, they hinge on the promises of technology. History tells us that, in the long run, coercion alone will never suffice to maintain a power structure, rendering magic superfluous. The technological infrastructure accumulated in certain areas of the world unevenly illuminates nightly satellite images. For the operation of the current global order to continue, it is no doubt essential that the billions of people whose daily labour maintains the asymmetric flows of energy and matter to these areas do not recognize, in the objects composing that infrastructure, the products of their own life force.

Consumption as a transformation of cannibalism

Let us conclude by pursuing some additional implications of this cultural analysis of capitalism. Graeber (2007:57–84) asks why our concept of 'consumption' builds on the metaphor of eating. He provides several persuasive historical hypotheses for why this metaphor is now applied to whatever people do when they are not working, including the urge to destroy things in order to gain recognition of one's sovereignty over them. Eating is indeed the perfect idiom for destroying something while literally incorporating it. But Graeber argues that many activities conventionally classified as consumption, such as watching television, do not involve goods that are destroyed by use. Nor, for the same reason, does he think that a teenage band practicing in a garage should be thought of as engaged in consumption. Yet, even these activities must submit to the twin constraints of capitalism and the Entropy Law (Georgescu-Roegen 1971) that correctly identify consumption as destruction: any activity that, for want of other resources, must involve manufactured goods – or even using electricity – implies destroying purchased physical resources in the process of creating meaning. The concept of consumption thus deserves to be retained for its critical potential: because it highlights how that which capitalism would have us maximize is ultimately destroying the planet. While there is no exemption from entropy – whatever the mode of production – the specificity of capitalism lies in its relentless pursuit of ever higher rates of resource destruction.

Graeber (2007) criticizes the concept of consumption from two opposite angles, namely for being perceived respectively as creativity and destruction. As much as I share his scepticism regarding the ideological uses of the former perception, I am unable to abandon the latter (even when applied to TV programming). In fact, it is only by acknowledging the material, biophysical dimension of the global economy that we can resist the seductive neoliberal glorification of consumption as the right to creative self-expression.

Marshall Sahlins' (1976) useful elaboration of Baudrillard (1972) taught anthropologists to view commodities as elements of semiotic systems that shoppers sought to incorporate into their selves, as the consummation of culturally constituted desires. Graeber (2007:57–84) traces the historical recognition that such consumer desires are potentially infinite and quite possible to manipulate. Clearly, it is this latter dilemma that raises the most incisive doubts about capitalism, rather than the extent of resource destruction itself. For if profits are proportional to our creative destruction of resources, it means that marketing will be geared to fabricating increasingly arbitrary incentives for us to maximize such destruction. To continue to expose this fundamental logic seems a more trenchant criticism of neoliberalism than to debate whether this or that activity is really destructive of resources.

Moreover, if the commodities we consume (i.e. metaphorically eat) are really em*bodi*ments of other people's life energy, not only is capitalism a transformation of slavery, as Graeber has argued, but of cannibalism.[10] The defining feature of capitalism is its specific social and cultural organization of the appropriation of geographically remote labour and land. Modern forms of market exchange, technology, and consumption represent net transfers of embodied (human) time and (natural) space extracted from some social groups for the disposal of others. Rather than directly controlling the labour of other human bodies in the vicinity, as in slavery, this is achieved by controlling the products of labour. Rather than shipping commodified labour (in chains) across the oceans, modern ocean-liners thus ship the commodified embodiments of labour. Ever since the first textile factories emerged in early industrial Britain, machines have assumed an illusory dissociation from the social relations of exchange through which their raw materials are extracted, appropriated, transformed, and redistributed. As I have argued in this chapter, this illusion rests on the cultural assumption that material objects are politically innocent and immune to moral critique. The same, ultimately Cartesian illusion liberates consumers to continue devouring distantly derived objects without any significant moral qualms about the social or ecological implications of consumption. As the use of general-purpose money and objectified market exchange were understood as immediate reciprocation and the severance of further social relations between market actors, the spirit of the gift was increasingly overshadowed by commodity fetishism. This cultural framework became solidly entrenched in the currently hegemonic economic discourse that was devised by (and for) successful stockbrokers such as David Ricardo (cf. Gudeman 1986), situated in the hub of a global empire.

A central paradox of this framework is that its point of departure appears to be a generalized power over objects, as exemplified by both consumption and by the fundamental severance of moral relations to an objectified environment (discussed above), while it simultaneously implies an unprecedented submission to objects, as exposed in Marxian analyses of fetishism. Although a prerequisite for modernity appears to have been an abandonment of animism, this very objectification may have paved the way for increasingly opaque varieties of fetishism. Paradoxically, to expose the political constitution of the machine would be tantamount to perceiving it as a malicious agent, i.e. to animating it, and perhaps to risk revitalizing the Luddite urge to subject it to physical sabotage.

I have tried to show that the concept of fetishism continues to be useful, not only within fields concerned with theology or the phenomenologies of aesthetic experience, but also for extending a general Marxian understanding of political economy. In particular, it can help us solve a neglected but puzzling conundrum of social science, namely how access to technological objects (i.e. 'development') can simultaneously be conceived as a result of exploitative accumulation and as the politically benevolent emancipation of all humankind. The answer, in this analysis, is that the fetishism of technology represents a specific mode of mystifying unequal exchange.

To sum up this chapter, we can ask ourselves what is the common denominator of the ideological pillars of modern power, which maintain the illusion of a morally neutral economy and technology, mystifying the affinities between capitalism and slavery, technology and magic, and consumption and cannibalism. I think the key is the phenomenon of *denial.* Johannes Fabian (1983) observed that the whole idea of development is founded on the denial of coevalness. The implicit assumption is that the people who don't own machines somehow inhabit an earlier period of time than those who do. In a similar manner, I suggest, the idea of the world market rests on a denial of appropriation. The concept of unequal exchange simply does not exist in the vocabulary of mainstream economics. Finally, our image of technology – much like commodity fetishism – is based on the denial of embodiment. In our Cartesian world-view, objects are automatically exempt from moral critique. And the denial of our co-existence with, exploitation, and consumption of other people is, like the Cartesian matrix as a whole, ultimately a dissociation from the reality of the Other.

3 Historical political ecology

Progress as environmental load displacement

The analyses of unequal exchange and machine fetishism presented in the two previous chapters have been influenced by world-system perspectives on international trade and accumulation (e.g. Frank 1967, 1978, 1998; Wallerstein 1974–1989; Braudel 1992 [1979]). In this chapter, I would like to outline some implications that such perspectives should have for our way of writing the history of environmental problems. There are different ways of writing such histories. The dominant mode has been to assume a common human history, a global 'we' experiencing the arrow of time through cumulative changes such as population growth, technological development, and new patterns of resource use. Ecological degradation, seen from this perspective, is the collective concern of a generalized humanity prompted to exploit new territories, harness new energy sources, and develop new transport technologies. Environmental problems, although alarming, are presented as the inevitable side effects of 'our' global success story.

The purpose of this chapter is to explore another way of writing environmental history. Rather than focus on the abstract accretion of landscape changes or technological inventions as a collective human experience over time, it seeks to highlight how such changes are distributed in space. It acknowledges that humanity is not a single 'we' but deeply divided in terms of reaping the benefits versus carrying the burdens of development. This *distributed* aspect of environmental problems can be illuminated by approaching ecological changes from a world-system perspective, viewing the world as a social system much more inclusive than individual nations (Wallerstein 1974–1989). If the world system for a long time has built on unequal power relations between rich core areas and impoverished peripheries, this inequality can also be expected to show in how environmental burdens have been distributed.

The political ecology of world systems

Over the past few decades, the concept of political ecology has become a useful shorthand for the growing recognition, in several disciplines, of the extent to which environmental changes and societal processes are intertwined. The many proponents of this approach, for instance in geography (Peet and Watts 1996; Bryant and Bailey 1997), anthropology (Paulson and Gezon 2005), and the philosophy of environmental justice (Low and Gleeson 1998), focus on what Martinez-Alier (2002:70) calls 'ecological distribution conflicts' in modern, often Third World settings. In integrating cultural, political, economic, and ecological perspectives on conflicts of interest between different social groups, political ecology requires transdisciplinary analyses that are able to handle the great variety of factors that enter into any such socio-environmental conflict. Generally driven by a deep concern with the predicament of marginalized, local people in today's world, it has made great progress in understanding the political economy of contemporary environmental change.

Although it is reasonable to assume that processes of environmental change were no less politicized in the past, studies of environmental history are much less frequently couched in such a political framework. The reasons for this are not hard to imagine. For one thing, environmental injustices seem less worthy of attention, and more difficult to document, when the victims are long gone. Nevertheless, there are many good reasons to apply the insights of political ecology to environmental history. Foremost, I would argue, is the importance of writing a politically valid environmental history of human civilizations, from the earliest agrarian empires to the present, industrial world order. The political ecology of human civilizations would no doubt present a completely different picture of 'our' global success story than mainstream historiography. In acknowledging the power inequalities and distribution conflicts framing the development of industrial civilization, it would ultimately provide a more realistic view of future prospects for sustainable development at the global level.

It is important here to emphasize that what I am suggesting is not merely an environmental history that acknowledges ecological degradation as the flip side of economic progress. This would be very far from a new idea. As mentioned in Chapter 1, Joan Martinez-Alier (1987) has traced the modern history of ecological economics, emphasizing how human economies are ultimately constrained by ecological conditions, to the mid-nineteenth century, and some of its roots to the eighteenth-century Physiocrats. As Donald Hughes (1994) has shown, such connections were obvious even in ancient Rome. Awareness of environmental limitations has no doubt always been a part of human experience, whether hunter-gatherers' concern over game stocks or horticulturalists' concern over soil fertility. Much recent work in environmental history in fact remains analytically confined to such Malthusian worries about a fundamental contradiction between human society and the natural environment. Irrespective of their discipline of origin – e.g. whether geography, archaeology, or history – most authors of classical works on global environmental history, even when addressing European imperialism, thus tend to couch their concerns in the generalized 'we' mode mentioned above (e.g. Thomas 1956; Turner *et al.* 1990; Simmons 1993b; Redman 1999; Diamond 2005). Rather than subject the internal structures of societal systems to critical scrutiny, such studies seem to assume a simple dualism of society versus nature and to account for environmental problems in terms of the inexorable progression of technology or demography.

Rarer are historical studies that explicitly investigate contradictions *within* global human society *over* the natural environment. These studies are generally concerned with European strategies of environmental load displacement during the sixteenth to nineteenth centuries and sometimes make explicit use of the world-system terminology (Wilkinson 1973; Wolf 1982; Mintz 1985; Worster 1988; Ponting 1991; Pomeranz 2000; Richards 2003). They demonstrate a more profound understanding of the societal dynamics responsible for unevenly distributed environmental degradation and often present direct continuities with studies of twentieth-century political ecology (cf. Bunker 1985; Goldfrank *et al.* 1999; Bunker and Ciccantell 2005). Yet, I am convinced that our understanding of how societal and ecological systems are intertwined can progress even further than this, not only by politicizing the environment, but also by ecologizing the world system.

It is frequently observed that globalization is not a new phenomenon in world history. Processes leading to greater societal connectedness over vast areas of the world can be traced centuries and even millennia back in time (Gills and Thompson 2006). A commonly neglected aspect of such processes is the inclination of more highly 'developed' regions – centres of civilization – to establish asymmetric relations of exchange with their local or global hinterlands, leading to accumulation of resources in the centres and often to environmental deterioration in their peripheries. This chapter briefly reviews some of the evidence for various kinds of such environmental load displacement in different societies and cultures over the past few millennia. Rewriting global environmental history in these terms allows us to conceptualize contemporary concerns with sustainable development and environmental security in analytically more precise ways, integrating perspectives from social and natural science and illuminating continuities and discontinuities between successive cultural patterns of resource use and their accompanying cosmologies.

Environmental historians have generally categorized their data in terms of a set of environmental problems that have been generated by specific historical forms of human resource use, production, consumption, settlement patterns, and so on. Some have produced comprehensive and authoritative accounts of the history of *one* such problem, as in studies of the history of energy use (e.g. Debeir *et al.* 1991 [1986]; Smil 1994), or in Williams' (2003) volume on global deforestation. Others have tried to sketch world histories of a whole spectrum of environmental problems. Thus, in a wide-ranging and popular compilation, Ponting (1991) reviews the environmental impacts of agriculture, deforestation, hunting, fishing, urbanization, pollution, energy use, etc. in different historical periods from the Neolithic to the present. Rarely, however, are environmental problems discussed specifically from the perspective of how they are distributed between different groups or sectors of society. Although such questions are often raised in modern discussions of political ecology and environmental justice, there has been little systematic or theoretical treatment of the uneven social distribution of environmental burdens in world history. When geographical distances between points of consumption and environmental impacts are addressed in historical studies, it is generally only to illustrate the widening spatial scale of environmental change, rather than to highlight issues of societal distribution. My goal in this chapter is to approach the uneven distribution of environmental burdens as a social strategy that is integral to the political economy of world-systemic processes, rather than the incidental effect of certain patterns of production and consumption.

All biological processes have 'entropic costs' (cf. Clark 1997:8–10) in the sense that they dissipate energy and materials. The earliest human societies did not differ much from our pre-hominid ancestors in terms of how these dissipative processes were organized. Local populations of hunter-gatherers derived their energy and material requirements from their immediate environment. Certainly there was both intergroup competition over resources and anthropogenic environmental impacts, but as long as neighbouring groups relied on the same kinds of resources, environmental degradation was no more of a *political* issue than among other species of mammals. The distribution of resources hinged on the ability of bands of foragers to gain and retain access to high-yielding territories, generally by defending their own presence in these areas. Social inequalities in resource distribution were primarily confined to local intra-group hierarchies based on age and gender. Technology was simple enough to rely almost exclusively on local resources and labour.

With cultural specialization geared to different ecological niches – such as coasts, floodplains, forests, and mountains – competition between human groups assumed new forms. Exchange of produce between different specialized zones introduced a new element in human–environmental relations: the possibility of systematically unequal transfers of embodied labour time or resources *between* communities or regions (cf. Godelier 1969). In line with what I have argued in the previous chapters, such unequal exchange can be conceptualized in terms of asymmetric flows of embodied land (gauged, for instance, in hectare yields) and/or embodied labour (gauged in human labour time) that have been invested in the traded products. Communities or regions that are able to sustain a net import of embodied land and/or labour are potentially able to invest the surplus in *capital*. Thus, capital *is* embodied land and labour. This relationship is well illustrated in Cronon's (1991) study, although not phrased in these terms, of how the city of Chicago emerged from the accumulation of inexpensive land and labour assets that had been invested in grain, lumber, and meat from the American West.

To reiterate a point made earlier, I use the concept of capital to denote some kind of infrastructure for locally saving time and/or space by intensifying the productivity of labour and/or land. If such infrastructures are also recognized as *constituted* by embodied labour and land that have been appropriated through unequal exchange, we are able to postulate the following relation between the three factors of production: capital is a strategy for locally saving labour (time) and land (space) through the *appropriation* of embodied labour (time) and land (space) from elsewhere in the social system. This understanding of capital carries very different implications to the conventional economic doctrine that posits land, labour, and capital as substitutable. In material terms, the three factors of production are *not* mutually convertible into one another. Land can nourish labour, and land and labour can yield capital, but capital can create neither land nor labour.

A brief history of environmental load displacement 4000 BC–AD 2000

Landesque capital (land improved for purposes of agricultural production) is the main form of capital accumulation in pre-industrial societies on all continents, and simultaneously one of the most tangible ways in which humans for millennia have changed their natural environments (cf. Widgren 2007a). There are many varieties – e.g. irrigation or drainage canals, terraces, raised fields, forest clearance, stone clearance, soil improvement – but one common incentive: to increase agricultural yield per unit of land. It is thus closely connected with sedentary settlement and often the object of violent competition or even military conquest. Widgren (ibid.) notes that the concept is in conflict with the notion of 'carrying capacity' as an immutable feature of a given environment. Although often attributed to central planning and hierarchical social organization, as in Wittfogel's classic discussion of Oriental despotism and 'hydraulic' civilizations, landesque capital can also be accumulated through the long-term labour investments of local communities, particularly under political, economic, and social conditions where investments in land are experienced as secure (ibid.). The concept of landesque capital reminds us that land and labour are not as distinct factors of production as they are sometimes visualized, for landesque capital is so by virtue of representing embodied labour. The great variation in the quality and productivity of land presents a great challenge for modern methods of natural resource accounting based on units of land, e.g. that of ecological footprints.

The landesque capital of early agricultural societies was often accumulated by chiefs or monarchs through the appropriation of labour by means of slavery, a labour tax, kinship, or other obligations. Fertilizers were sometimes transported over considerable distances. Thus, the local intensification of agricultural production was frequently made possible through the mobilization of human and natural resources from a much wider area. Local increases in productivity at times implied ecological and/or human impoverishment elsewhere. The advantages of agricultural intensification vis-à-vis foraging, however, were not so much a matter of saving time as of saving space. It made it possible to concentrate larger populations in restricted areas, a project that primarily served the interests of political-religious elites who sought control over ever larger workforces. For the majority of the population it implied more and harder work, less freedom, and poorer health (Hughes 2001:25, 32). Although population growth is often referred to as a driving force behind agricultural intensification (cf. Boserup 1988), it should be recognized that local demographic expansion need not primarily be a biological phenomenon, but can result from centripetal socio-political processes driven by elite strategies to accumulate rising numbers of servants, artisans, and soldiers.

If landesque capital refers to various inalienable investments augmenting the productivity of land, we can also identify a specific kind of capital designed to assure the cooperation of labour. In order to be able to mobilize and control large populations, the rulers of early theocratic civilizations needed to invest in ritual capital such as prestige goods, temples, palaces, and other public architecture. The production of such capital generally relied on resources far beyond the city limits. In the fourth millennium BC, for instance, the uplands to the north and east of Mesopotamia suffered from deforestation (and subsequent soil erosion) due to timber and charcoal extraction catering to the demands of public architecture in lowland centres like Uruk (Redman 1999:180–182; Hughes 2001:36–37). Sumerian merchants also brought timber, ivory, and precious stones from as far as the Indus Valley, and copper from Cyprus. In the Indus Valley civilization (2300–1800 BC), mud bricks were dried in ovens that required great quantities of wood, and the resulting deforestation caused rapid soil erosion over a wide area. Early Indus Valley cities also imported cedar wood from the Himalayas, while Sumer, Egypt, and Rome harvested cedars and other timber from Lebanon. Most of the Mediterranean area was deforested to provide grain, fuel, and construction materials for the urban centres of ancient Greece and Rome. Roman ships brought tin and other scarce resources from beyond the Strait of Gibraltar. Other, more local examples of deforestation and ecological deterioration largely due to the excessive construction of elite monuments include Easter Island and the classic Maya (Hughes 2001:45–46). Such ritual capital, the rationale of which is primarily symbolic, should be acknowledged not as superfluous luxuries but as an integral aspect of the technology of early state power. No less than other forms of capital, it has generated tangible material repercussions, locally as well as throughout its hinterland.

Exports of foodstuffs and other perishable goods such as wool are difficult to trace archaeologically, but are likely to have occurred in the Near East already around 7000 BC, in exchange for archaeologically documented imports such as obsidian, turquoise, and copper. The urban core area of lowland Mesopotamia in the fourth millennium BC increased its export production of grain, ceramics, and textiles in order to sustain imports of timber, minerals, and other resources not locally available (Redman 1999:185–186). Agricultural intensification in the core area led to shortening of fallows, cultivation of marginal lands, and salinization. Meanwhile, the extraction of timber, stone, and metals in the peripheral uplands east of Mesopotamia also led to environmental degradation. Ancient Egypt exported grain, other foodstuffs, flax, and papyrus, and became the breadbasket of the Roman Empire. Beginning in the third millennium BC, an increasing export production of olives and grapes around the Mediterranean reduced natural forest cover and transformed the landscape. In the Levant, these environmental impacts peaked during Roman colonial rule about two thousand years ago. Such environmental changes are reflected in the successive extinction of species such as elephants, rhinos, hippos, giraffes, and zebras in Egypt and North Africa, lions and leopards in Greece, and tigers in Mesopotamia (Ponting 1991:161–162).

Deforestation in Europe reached a peak by 1300, although the bubonic plague was followed by two centuries of temporary regeneration (Hughes 2001:86, 92). A few decades earlier, China experienced a similar episode of demographic collapse and temporary environmental recovery. At this time, Italian cities like Florence, Venice, and Genoa were drawing on resources from all over Europe and the Mediterranean basin, including wool from an increasingly deforested England (ibid., 87–89). By the early fourteenth century, Florence had to rely on long-distance imports even of food, but these supplies were insecure and famine remained a real threat. The shortage of wood became increasingly severe in Europe as consumption increased in architecture, fortifications, heating, lighting, metallurgy, and the manufacture of glass, tile, bricks, pottery, carts, ships, weapons, musical instruments, dishes, shoes, barrels, and other products (ibid., 90). The production of one ton of iron, for example, required the annual growth of twelve hectares of productive forest. Hansa ports such as Lübeck controlled a lucrative trade in timber and other forest products, such as furs, from around the Baltic. By the sixteenth century, most Portuguese ships were built in India or Brazil, while Spain imported timber from Poland in order to build the Armada, and England imported timber from Scandinavia and Russia (Ponting 1991:278–279). Already by the late seventeenth century, British warships were being built in North America.

Agricultural intensification is customarily presented as a process of technological improvement designed to feed greater numbers of people, but as Ponting (1991:108–109) reminds us, famine is generally not so much a matter of food shortage in absolute terms as of who is given access to food. In recent centuries, this has largely been a matter of who can afford to buy it. Intensification tends to be geared to export production, and food exports often continue from areas suffering from famine. This is particularly evident when agricultural intensification occurs in economically peripheral areas. Historical examples include fourteenth- and fifteenth-century France, nineteenth-century Scotland and Ireland, India in 1876–1877, Russia in 1911–1912, Bengal in 1943, and more recent famines in Ethiopia and the Sahel. In all these cases, it is apt to speak of a displacement of environmental burdens from those regions where the exported food is consumed to the impoverished areas where it is produced. After about 1850, Europe had decisively shifted to such a displacement strategy through large-scale imports of food and fertilizers from its colonial periphery. Although this periphery – today known as the South – suffers recurrent famine and widespread malnutrition, it has long continued to export more food than it imports (ibid., 214). The most recent surge of agricultural intensification – the so-called Green Revolution in the economic periphery of the world system – has had profoundly detrimental socio-political implications in terms of redistributing resources and environmental burdens at both the local and global level (cf. Hughes 2001:186–187).

The four centuries of European expansion between 1500 and 1900 have aptly been characterized as 'the rape of the world' (Ponting 1991:161–193).

Not only was this expansion – like those of previous civilizations – founded on slavery and other forms of forced labour, but the relentless pursuit of profitable resources from other continents wrought havoc on distant ecosystems and societies in order to satisfy European desires for, among other things, precious metals, ores, timber, plantation crops (e.g. sugar, tobacco, cotton, rice, tea, coffee, bananas, rubber, cocoa, palm oil), fertilizers, fish, whale oil, ivory, furs, and feathers (cf. Richards 2003). The most tangible environmental consequences of this global expansion of resource extraction include devastated mining landscapes, deforestation, soil exhaustion, collapsed fisheries, and a dramatic reduction in biodiversity. The specialized cultivation of cash crops for the international market not only exhausted the soil but also reduced local food security in a number of countries, including Cuba, Fiji, Tahiti, Gambia, and the Philippines, increasing their reliance on food imports (Ponting 1991:213). A very tangible example of how European accumulation brought impoverishment to its colonial periphery is the import of fertilizers such as guano from South America and phosphates from Nauru and other islands in the Pacific Ocean (ibid., 218–221; Clark and Foster 2009). Today, affluent countries are able to maintain fertile soils and productive forests by importing resources (e.g. oil, nitrogen, phosphorus) to alleviate pressure on the land, a development strategy that can hardly be offered as an example for less affluent nations to follow.

The rise of European hegemony in the late eighteenth and nineteenth centuries was founded on new possibilities of environmental load displacement provided by access to the New World and investments in industrial technology (Wilkinson 1973; Pomeranz 2000). Industrial capital (i.e. machinery) can be distinguished from landesque and ritual capital in that, rather than merely improving the productivity of land or cooperativeness of labour, it can *replace* labour in harvesting and refining the products of the land. In processing raw materials imported from elsewhere, moreover, it can locally also serve to replace land itself as a productive asset. To be sure, the three forms of capital are continuously interwoven with one another, but this analytical distinction remains valid. What is not generally recognized, however, is that the substitutability of industrial capital for labour and land is more apparent than real (Hornborg 2001b). From a local perspective, for example England in the nineteenth century, capital may seem to substitute for labour and land, but this illusion stems from a failure to grasp the logic of the entire global system from which this capital has been generated. The accumulation of capital in nineteenth-century England represented vast quantities of embodied labour and land appropriated on other continents (cf. Chapter 5).

The twentieth century brought new and even more dramatic technologies of environmental load displacement, through which 'developed' regions were able to shift much of the industrial degradation of soil, air, water, and vegetation to their 'developing' peripheries. The most obvious innovation of the twentieth century was the massive increase in combustion of oil, which has generated environmental problems at every moment of processing from extraction and transport through consumption, garbage disposal, air pollution, and climate change. Whereas 'developed' regions such as North America, Europe, and Japan have the highest rates of per capita consumption of oil, it is the 'developing' regions that are hardest afflicted by the environmental consequences. Assuming that these consequences (e.g. floods, droughts, hurricanes) can be estimated in terms of economic externalities, Martinez-Alier (2002:213–233) has calculated that the entire external debt of Latin America in 1999 (US$700 billion) would be cancelled by a mere twelve years of the United States' 'carbon debt' (US$60 billion/year). Taking this ecological debt seriously would thus mean acknowledging that Latin America's foreign debt is a mystification of actual debtor–creditor relations.

An analytical framework for understanding environmental injustice

This brief sketch of the history of environmental load displacement from ancient Mesopotamia to current concerns about global warming can provide us with the outline of an analytical framework for understanding the uneven distribution of environmental problems as a world-systemic process. Rather than lump all instances of environmental deterioration into a general category of ecological consequences of civilization (cf. Chew 2001; Diamond 2005), it is important to distinguish between the different structural conditions that in separate historical periods generate varieties of environmental change. Apparently, it is still possible to write best-selling volumes on the history of environmental problems without any reference to social science (Diamond 2005). From a social science perspective, however, it is crucial to specify the social units of analysis experiencing environmental deterioration and whether this deterioration is primarily the result of endogenous or exogenous factors.

What Clark (1997:8–10) calls the 'entropic costs' of industrial metropoles should be clarified as comprising two dimensions: (1) the appropriation of *negative entropy* (accessible, high-quality energy and materials) from elsewhere and (2) the displacement of *entropy* (energetic and material disorder) to some unspecified hinterland (cf. Frank 2007). In the present world order, the first could be exemplified by the struggle of industrial powers to control vital sources of fossil fuels (Klare 2001), the second by controversies over various kinds of ecological distribution conflicts (Martinez-Alier 2002:258–260), best illustrated by the uneven use of sinks for carbon dioxide and the uneven share of the burden of its climatic consequences (ibid., 213–233; Roberts and Parks 2007, 2009).

One of the benefits of this analytical framework is that it would help us to distinguish between different kinds of environmental problems in world-system history. In view of the twin problems of importing available energy (negative entropy) and exporting entropy, different historical and contemporary cases of environmental crisis require different analytical tools. There are cases, such as the Classic Maya or Easter Island, where ecological overshoot is not so much a consequence of long-distance resource extraction as of locally generated overexploitation of resources. There are clearly also a great number of cases, such as the hinterlands of Sumer, Roman North Africa, thirteenth-century England, nineteenth-century North America, or twentieth-century Amazonia, where environmental degradation was the result of the systematic appropriation of local resources by distant centres. A third type of environmental problems would be represented by nineteenth-century London or twentieth-century Moscow, where the accumulation of distantly derived resources was not sufficiently balanced by an export of the entropy (pollution, waste) generated in the process. Finally, the recent North-to-South export of the most polluting industries, as well as waste itself, suggests a fourth version of ecological crisis, where world-system peripheries are converted into dumping grounds for entropy generated by affluent core areas (cf. Eisenmenger and Giljum 2007; Muradian and Giljum 2007). This phenomenon, epitomized in 1991 by the so-called Lawrence Summers Memo, is particularly conspicuous around North–South boundaries such as the border between Mexico and the United States (cf. Heyman 2007). These different varieties of environmental degradation clearly illustrate that problems of environmental justice and environmental load displacement can have two distinct aspects, since peripheral areas can be exploited both as sources of negative entropy and as sinks for entropy.

It seems that civilization as we know it has generally been founded on processes of environmental load displacement and great disparities between affluent core areas and impoverished peripheries. The hinterlands of ancient Mesopotamia, the Indus Valley, Egypt, Greece, and Rome were all afflicted by deforestation, erosion, mining, and loss of biodiversity that ultimately contributed to the decline of the core areas. During recent centuries, emerging from the assets accumulated through European colonialism, the expansion of industrial capitalism has entailed more sophisticated and less obvious strategies of environmental load displacement, to the detriment of the global South. Current concerns with sustainable development and environmental security need to more explicitly address the political issues of how environmental problems are distributed in global society. To frame the discussion in terms of whether 'we' shall 'fail' or 'succeed' in achieving sustainability (Diamond 2005:3) is to mystify the underlying political contradictions and represent environmental security as an uncontroversial managerial challenge for an unspecified social unit of analysis. When civilizations succeed – which they all do for a time – they generally tend to do so precisely by exporting their environmental problems, and jeopardizing sustainability elsewhere. When civilizations fail – which they all ultimately do – core and periphery alike tend to suffer through the subsequent turmoil, but former peripheries are the more likely to emerge as future cores (cf. the so-called 'law of evolutionary potential', articulated by Marshall Sahlins and Elman Service already in 1960).

Approaches to ecologically unequal exchange

A fundamental theoretical question, for any attempt to merge ecology and world-system analysis, is whether the global distribution of environmental deterioration is somehow structurally determined. If the answer is yes, does this structural determination hinge on the physical properties of the traded commodities? Are economic sectors specialized in the extraction of natural resources systematically disadvantaged – economically and/or ecologically – vis-à-vis sectors specialized in the production of industrial goods? In this section I will review some theoretical and methodological approaches to this issue. These contributions generally propose to deal only with modern (capitalist) economic systems, but in the final section we shall consider the extent to which their analytical frameworks are applicable also to pre-modern societies.

Martinez-Alier (1987) has traced the early dialogue between Marxism and natural sciences such as physics and agricultural chemistry on the material dimensions of economic productivity, illustrating the difficulties that continue to plague communication between the social and natural sciences on economic matters. Although Marx was keenly interested in the metabolic flows between urban and rural areas, he did not think of these flows as inherently exploitative or as a constraint on progress, as did Malthus, but was optimistic about new technological remedies such as chemical fertilizers. Martinez-Alier suggests that if Marx and Engels had paid more attention to Serhii Podolinsky's argument for an integration of thermodynamics and their notions of surplus exploitation, early Marxist theory might have incorporated the kind of socio-metabolic perspectives that today provide a foundation for discussions on ecologically unequal exchange.[1] It is reasonable to assume that the technological optimism of traditional Marxism has been geared to the political agency of the labour movement in industrial capitalism, whereas serious and systematic ecological critiques of capitalism date from the late twentieth century (e.g. Martinez-Alier 1987; O'Connor 1994; Foster 1999).

Stephen Bunker (1985) pioneered the integration of ecology and world-system analysis. He posited a structural asymmetry between extractive economies in the periphery and 'productive' economies in the core, which has been exacerbated with the increasing spatial separation of extraction and production. The structural polarizations of production versus extraction and of core versus periphery are in fact one and the same. One of his key points is that the dynamics of scale have inverse consequences for productive versus extractive sectors, yielding falling unit costs of production in the former but rising costs in the latter. Furthermore, as resources are depleted or technologies and market demand transformed, extractive economies are unable to sustain the continuous, cumulative development of labour and infrastructure that, by and large, have characterized productive core areas over the past two centuries. To account for such uneven development under capitalism, Bunker suggested that the Marxian labour theory of value ought to be complemented with a notion of 'natural values', which, like labour, are systematically underpaid by the industrial core areas to which they are transferred. More recently, Mario Pérez Rincón (2006) has argued for an 'ecological Prebisch thesis', i.e. a structural understanding of development gaps that similarly accounts for economic disparities in terms of the specialization of different regions in extraction versus production.

The economists Roldan Muradian and Stefan Giljum (2007) have applied the concepts and methods of material flow analysis to investigate how environmental burdens are distributed between nations. Their first finding is that so-called polluting sectors have generally increased their physical volume of export production in absolute terms over the past few decades, irrespective of world region. Second, they find that, in the period between 1978 and 1996, the proportion of total exports that derive from polluting sectors, measured in monetary value, has increased in South America and Africa while decreasing in Europe, the United States, and Japan. Third, they note that the European Union has remained a net importer of products from polluting sectors during the whole period analysed.

These findings are highly relevant for assessing the economists' notion of *dematerialization* – i.e. the proposition that economies will become less of a burden on the environment the more developed they are – and the sobering counter-proposition known as the pollution haven hypothesis – i.e. that polluting industries tend to migrate toward poorer countries with weaker environmental standards. Much as the economist Lawrence Summers infamously suggested in 1991, there is an impeccable economic logic in advocating a migration of dirty (i.e. polluting) industries to less developed countries. Although the appearance of a cleaner, seemingly dematerialized economy in affluent nations may suggest that economic growth is good for the environment, such a view may be the product of an illusion. An alternative interpretation is that economic growth makes it possible to shift a country's environmental burden to other, less affluent countries. Moreover, we have reasons to be sceptical of indices of sustainability that measure environmental burdens relative to GNP, suggesting that it does not matter if environmental loads increase in absolute terms, as long as GNP increases faster. The illusion of dematerialization thus simultaneously highlights the two main obstacles to understanding socio-environmental metabolism using the kinds of data conventionally available to economists: the fact that statistics on trade generally use the nation as unit of analysis, and the fact that these statistics generally use monetary value as the only significant metric.

How nations constrain our thinking about unequal exchange

The significance of the nation-state in organizing socio-environmental metabolism is undisputable, but in different ways than economists and historians have generally recognized. In the age of globalization, it should no longer be possible to imagine that a nation's geographical extent, its economic activities, and its environmental impacts coincide. Yet, to correlate national statistics on GNP and environmental quality in order to draw conclusions on the relation between economic growth and environmental performance is precisely to make such an unreasonable assumption. From a world-system perspective, on the contrary, it is only to be expected that one nation's environmental problems may be the flip side of another nation's affluence. Such connections are evident even where there are no indications of coercion or imperial ambitions.

What, then, is the role of national boundaries in organizing the distribution of environmental resources and risks? They are obviously instruments of political agency, even when there are no attempts to expand them, simply by defining arenas within which power can be asserted in direct and uncontroversial ways. They may also define fields of responsibility and a minimum of social solidarity, prompting governments – to the extent that they are able and willing – to redistribute resources so as to cushion marginalized groups within their jurisdiction against the worst ravages of the polarizing tendencies of accumulative processes. Rural populations engaged in extractive activities within affluent nations can thus be financially subsidized by successful industrial sectors of the same nations, making it possible for them to enjoy high standards of living, healthy environments, and security, whereas their counterparts in poorer nations may be left virtually unprotected. The nation-state is thus the basic social unit capable of politically intervening in the socio-environmental, metabolic processes generated by the logic of capital accumulation. Not only can it decide to alleviate the impoverishment of its internal, national periphery; ever since the days of mercantilism, states have explicitly struggled to promote – as national interests – activities that serve to make their external positions within world systems as advantageous as possible. For centuries, the more powerful nation-states have used trade policies backed up by military force to secure vital flows of resources from the outside. Not so long ago, in fact, nation-states regularly struggled to expand their boundaries to achieve such goals. As trade has replaced conquest and tribute, national boundaries have begun to appear more stable over time, but the use of military force to secure resource flows is far from obsolete. Increasingly, states have also made it a national interest to export environmental risks.

As anthropologist Joe Heyman (2007) has shown, this gate-keeping function of nation-states, struggling to import resources and export risks, is most obvious on their borders. Focusing on the border between the United States and Mexico, he explores the many subtle ways in which this boundary zone serves to maintain the difference between citizens belonging to the two adjacent polities. This serves to remind us that the formation of nation-states involves a partly arbitrary construction of territorialized identity, but also that territorializations of socio-metabolic relationships have very tangible consequences for people who happen to be living on either side of the border. The polarizing structures of world-systemic processes of capital accumulation are not necessarily organized in spatial terms to begin with, but the medium of nation-states has historically projected such structures onto geographical territories. It remains to be seen how far the current surge of globalization can go in challenging our attachment to the nation-state as a source of identity and a category for understanding the structure of world society. Heyman (ibid.) illustrates the very real, material processes that continue to undergird the existence of nation-states precisely by asserting their boundaries. The environmental dimension of this particular boundary is epitomized by the *maquiladoras*, a highly polluting industry concentrated along the Mexican side of the border, where wages can be kept low and environmental considerations ignored, but the profits from which often wind up among owners and executives residing north of the border. It is difficult to think of a more straightforward verification of the pollution haven hypothesis.

The exploitation of the periphery is not a myth

The metabolic rift between economic cores and their extractive peripheries can be identified from ancient Rome to nineteenth-century England, and the tendency to displace environmental burdens ever further off from core areas is pervasive, irrespective of whether it implies extending or transgressing political boundaries. The ecological impacts of long-distance trade in the ancient world can be exemplified by the Roman appropriation of timber, grain, and metals from vast areas of Europe, Africa, and Asia. From the sixteenth century and on, such impacts are all the more obvious, e.g. in the extraction of timber and grain from eastern Europe, silver from the Andes, ivory from East Africa, and sugar from the Caribbean (cf. Håkansson 2007; Moore 2007; Williams 2007). The uncalculated, indirect environmental impacts of the extraction of e.g. silver, ivory, or sugar are important aspects of the ecological trade balances of colonial Spain and Britain. The modern displacement of the most polluting industries from the European Union to Latin America thus has antecedents in sixteenth-century silver mining. Loss of biodiversity in the periphery is a theme that can be traced from Roman North Africa to the savannas of modern Brazil (cf. Ponting 1991:161–193; Richards 2003:463–616; Fisher 2007). Another pivotal factor in human–environmental relations throughout the millennia is the accumulation, maintenance, and eventual abandonment of landesque capital, designed to produce e.g. grain, wine, olive oil, rice, silk, beef, sugar, cotton, or soya beans as commodities for impersonal, monetized markets. Equally ubiquitous is the lack of environmental foresight that characterizes most market-oriented extractive activities, whether conducted by Roman slaves or nineteenth-century loggers (Wilk 2007).

It is obvious that many of the imports to core areas, from ancient Rome to nineteenth-century Europe, were of negligible direct significance for the social metabolism of these areas. If we were to restrict our perspective to this observation, we might well agree with economic historian Paul Bairoch's (1993:97) remarkable conclusion that 'the West did not need the Third World'. But in looking at early twentieth-century statistics to draw the conclusion that the dependence of the West on raw materials from the Third World prior to 1955 is a 'complete myth' (ibid., 70), Bairoch seriously distorts our view of the dynamics of economic development and industrialization. Let us look closer at Bairoch's position as a foil against which to consolidate the argument in this chapter.

First, as Bairoch (ibid., 88–97) himself concedes, the consequences of incorporation as a world-system periphery can be destructive enough, even if the socio-metabolic significance of the exports for importing core countries seem insignificant. To use a modern expression, the ecological rucksacks of such exports (e.g. spices, jewels, silver, gold, silk, ivory, feathers, sugar, tobacco, coffee, tea) were generally formidable, and this is but one of the ways in which the periphery was – and continues to be – forced to pay a high price for the economic development of the core.

Second, Bairoch's conclusion that the role of colonialism was unimportant in the birth of the British Industrial Revolution (ibid., 80) underestimates the recursive relation between mercantile and financial profits, on one hand, and investments in mechanization, on the other (cf. Wolf 1982:267–295). His mechanistic, linear notion that 'during the eighteenth and nineteenth centuries colonization was primarily a result of industrial development and not vice versa' (Bairoch 1993:82) fails to appreciate the complex ways in which these two phenomena were mutually reinforcing aspects of the same process.

Third, his use of the category 'the West' as a static economic and/or political entity extending indefinitely backwards in time beyond 1955 is extremely misleading. Economic core-periphery relations are dynamic and multi-scaled and frequently shift over the course of history, not to mention the variously constructed political boundaries asserted by different governments so as to maximize advantages and minimize risks in particular historical contexts. The imperial projects of Rome, Spain, and Britain are obvious examples. A socio-metabolic perspective on the accumulation of industrial infrastructure must avoid being straitjacketed into the fetishized nation-state categories that organize economic statistics. We do not need to go very far back in time to find large parts of Europe peripheral to the industrial and commercial developments in Britain and the Netherlands. Nor should we forget how recently North America was a source of cheap (largely slave) labour and land serving as a vast extractive zone particularly for Britain. If Bairoch (ibid., 59–71) finds that, just prior to the First World War, the 'developed West' was basically self-sufficient in minerals and other raw materials, it certainly does not mean that the accumulation of industrial infrastructure within its present-day political boundaries has not historically been characterized by ecologically unequal relations of exchange between core regions and *what were then* their peripheries.

The fundamental structural relationship, if we are to understand the social metabolism underlying economic development, is the exchange between geographical spaces experiencing an accumulation of physical capital (cores), on one hand, and extractive areas suffering net exports of natural resources (peripheries), on the other. Political boundaries and national statistics frequently – if not systematically – distort and mystify this relationship. At certain times and places an industrial core region in a given country will find its domestic, national periphery more or less sufficient for its metabolic requirements, at other times not. Although national statistics would make them seem comparable entities, Canada and Singapore obviously have very different capacities in this respect. Although the basic structural imbalance, which Marx identified as the metabolic rift between town and countryside, is as old and pervasive as urbanization itself, the past few centuries of globalization have seen an increasing ambition – and capacity – of nation-states to displace such imbalances beyond their own borders to the international arena.

Finally, our review of the historical evidence from the past two millennia clearly shows that developments in core areas, thus defined, have in fact frequently involved imports of bulk commodities, the significance of which, although less voluminous per capita than today, was very far from negligible for their metabolism: foodstuffs, timber, metals, fuels, and fibres are only the most obvious examples. Even if the mid-twentieth century to Bairoch suggests a discontinuity in the sense that extraction of such resources for the 'developed West' (Europe and North America) was increasingly externalized beyond its modern political boundaries, there are clear historical continuities in the fundamental structures of ecologically unequal exchange. Not only was the 'developed West' a rather recently constituted geographical and political category at that time, but, over the centuries, structures of such unequal exchange have frequently unfolded with little regard to political boundaries. As long as it is confined to statistics on the flows of exchange values crossing national borders, the discourse on development and underdevelopment will thus be severely constrained.

4 Towards a truly global environmental history

Although rich in empirical detail, studies in environmental history often strike world-system analysts as theoretically underdeveloped. They generally do not address the fact that landscape changes in core areas have been recursively interconnected with those in peripheral areas. Although several recent books claim to deal with global environmental history (e.g. McNeill 2000; Hughes 2001; Richards 2003; Radkau 2008; Simmons 2008), they are rarely 'global' in this sense. They tend to offer a series of national and local case studies, focusing more on the environmental records of individual nations and groups than on the global historical processes and material flows that have generated their problems as well as their options. In terms of Pomeranz's (2000:25) useful distinction, most of these global narratives treat different regions in terms of 'comparisons' rather than 'connections'.

The main theoretical conclusion of much of this work is the recurrent observation that technologies designed to solve one kind of problem will ironically tend to generate even more severe problems of another kind, often for other groups of people. This conclusion should instil a powerful antidote against the pervasive belief in technological solutions, but the underlying message is generally that there is something inevitable and politically innocent about the course of global environmental history. Considering that many of these authors use words such as 'global' and 'world' in the titles of their books, it is remarkable that so few of them really consider the world as a system, in which environmental transformations in two geographically distant countries or regions may be closely intertwined in terms of causal connections. There is very little recognition of the fact that economic expansion in one area often implies environmental load displacement to other areas.

The first part of this chapter critically reviews some recent treatments of human–environmental relations over the long term by historians, archaeologists, and geographers. The second part elaborates a framework for understanding global environmental history based on a world-system perspective and on a concept of accumulation that acknowledges both the semiotic and the material dimensions of exchange in human societies.

Contributions from different disciplines

The expansive transdisciplinary project of 'environmental history' has united historians, geographers, archaeologists, anthropologists, sociologists, economists, ecologists, agronomists, foresters, and a host of other academic professions in tracing environmental transformations over time. This section reviews some recent contributions to this field – with vantage points from disciplines as diverse as economic history, archaeology, and biogeography – in terms of their more or less explicit analytical frameworks. It critically discusses their theoretical assumptions, for instance regarding a recurrent inclination toward Eurocentrism or European exceptionalism.

Environmental history: Hughes, Radkau, McNeill, Pomeranz, Richards, Ponting

In *An Environmental History of the World* (2001), Donald Hughes presents a number of case studies of human–environmental interaction arranged chronologically to illustrate eight 'general periods in human history … characterized by large-scale changes in the relationship of human societies to the biosphere' (ibid., 9). His ambition to use case studies to exemplify 'the larger picture' (ibid., 3) raises the question of what this larger picture finally comprises. Although he quotes Aldo Leopold on the imperative of fusing ecology with the findings of sociology, economics, and history (ibid., 6), Hughes' attempts to transcend the discipline of history lead only to ecology, never to any significant contributions from sociology or economics. His vision of environmental history is 'a world history that adopts ecological process as its organizing principle' (ibid., 8), leaving readers from the social sciences wondering if global social organization over the long term can really be reduced to ecology. In his wide-ranging book, Hughes briefly considers the role of long-distance demands on natural resources, as in the distant environmental impacts of cities in Mesopotamia, Rome, and medieval Italy, but these are momentary excursions that do not discuss such environmental load displacements as a recurrent phenomenon in world history. Like most other historians narrating global environmental change, Hughes does not seem to consider societies themselves as globally organized. Instead, the recurrent sources of environmental change are presented in the fragmented terms of factors such as population growth, new technologies, inadequate ideas and ethics, lack of foresight, and greed. As we shall see, the absence of insights regarding the global organization of society, from which such local phenomena can be derived, is currently pervasive in the narration of global environmental history.

In Joachim Radkau's book *Nature and Power: A Global History of the Environment* (2008; German edition 2002), the emphasis on the Old World, Europe in particular, provides a welcome complement to the widespread preoccupation with the Americas and Oceania that has dominated the first decades of environmental history (e.g. Crosby 1972, 1986; Worster 1979; Cronon 1983; Merchant 1989; Kirch and Hunt 1997; Balée and Erickson 2006). The volume illustrates how concerns with sustainable human–environmental relations have been central to European consciousness many centuries before the colonization of the New World. In rich anecdotal detail, Radkau's narrative shows how local populations and central administrators have dealt with recurrent problems of soil degradation, deforestation, irrigation, and the pollution of water and air. However, there is very little theoretical treatment of the historical data. Radkau's environmental history is not 'global' in the sense that it shows how environmental changes in different parts of the world are interconnected. It focuses on the environmental records of individual nations and (essentialized) religions and peoples – such as Muslims and Indians (or 'Indios', as he frequently calls them) – rather than on global historical processes. There is also a troubling inclination toward an explicit European exceptionalism, as in assertions that in Central and Western Europe 'environmental problems were more or less solved' (p. 151). Referring to, among others, the biogeographer Jared Diamond (whom he calls an 'ethnologist'), Radkau lists the features that supposedly made Europe uniquely sustainable: the many domesticated animals; the abundance of forest; the robust soils; the regularity of rainfall; the many streams suitable for water mills; the legal and political institutions; etc. (pp. 184–194). Where others (e.g. Wilkinson 1973; Pomeranz 2000) have seen European expansionism as a strategy of environmental load displacement, i.e. as a response to socio-ecological crisis, Radkau sees it as a sign of stability and success (p. 204).

Radkau's concern with 'power' seems almost completely restricted to the sphere of politics and policies, whereas serious critical analysis of the environmental implications of economic systems is just as absent as world-system analysis. Although occasionally critical of the 'concentration of capital' and the 'quest for private profit' (p. 299), he suggests that environmental protection is 'dependent on cooperation with economic forces' (p. 280) and that to draw adversarial battle lines between ecology and economy is 'absurd' (pp. 305, 313). He does note, however, that the new petroleum energy regime associated with the post-war '50s syndrome' (p. 252) represents a decisive discontinuity in human–environmental relations.

For Radkau, as for many other historians, environmental history seems primarily to be the history of environmentalism. His ambition to offer general statements on a phenomenon as heterogeneous as the pervasive concern with human–environmental relations can only lead to inconsistencies and contradictions, as in his discussions of local and individual agency vs. large-scale, central coordination, and in his ambiguous account of the German love for a nationalized *Heimat* (pp. 236, 308) or the 'exemplary' environmental record of the dictatorial 'ecotopia' of Bhutan (p. 285). Although frequently an advocate of local community wisdom and tacit practical knowledge, he has an unmistakable faith in the authority of the nation-state. Similarly ambivalent are his assessments of national vs. global environmental policy (pp. 286–288). His apparent determination to accommodate and legitimize perspectives ranging from sober rationality and pragmatism to romanticism and mysticism finally leaves the reader asking for a bottom line. No wonder that a German radio journalist, to the author's consternation, presented *Natur und Macht* as a dismissal of environmental worries (p. 326). There are a number of normative assertions, to be sure, but hardly any theoretical conclusions, except perhaps the recurrent observation that solutions tend to create new problems.

This seems precisely also to be the central conclusion of J. R. McNeill's celebrated volume *Something New Under the Sun* (2000), focusing on the environmental history of the twentieth century. McNeill has organized his chapters in terms of the biophysical character of the environmental problems that he considers, e.g. whether relating to soil, air, water, biodiversity, demography, or energy use. This makes his book more useful and easier to read than Radkau's, and it offers an impressive range of historical and quantitative data on global environmental change over the past century. Like Radkau, however, McNeill seems to have little ambition to offer a theoretical understanding of the societal drivers of landscape change. His account reflects extensive reading, but the analytical interpretations are based on common sense and generally do not transcend images of environmental history that most educated citizens already subscribe to. Yes, there has been an abundance of cheap energy (for some), economic growth (for some), population growth, urbanization, industrialization, air and water pollution, soil erosion, deforestation, and biodiversity loss. In addressing the driving forces behind these well-known changes, McNeill refers very briefly to 'human ingenuity' and 'luck' (pp. 16–17), or (in passing) to 'the attractions of higher living standards, or at least more consumption, for masses of people; of profit for hundreds of firms engaged in mining, metallurgy, electric power generation, and other polluting enterprises; and of political power for states, bureaucracies, and politicians' (p. 116).

The main theoretical insight that can be extracted from McNeill's volume is thus very similar to what we might conclude from reading Radkau, namely that technologies designed to solve one kind of problem will, ironically, tend to generate other and even more severe problems of another kind. This has, for instance, repeatedly proven true of energy use, new transport technologies, the industrialization of agriculture and fishing, refrigeration, irrigation, the introduction of new species, and mass production for mass consumption. The impression conveyed by these texts is that there is something inevitable about environmental degradation, at least in terms of the drivers of environmental problems, if not in terms of how politicians tend to deal with them. Neither Radkau nor McNeill in their grand syntheses discusses the fact that environmental transformations in two geographically distant regions may be closely interconnected, or that some regions and populations may become the impoverished and disempowered victims of accumulative strategies originating elsewhere.[1]

By contrast, in *The Great Divergence* (2000), Kenneth Pomeranz shows exactly this. He argues that the unique economic and technological developments in eighteenth- and nineteenth-century Europe should not be explained with reference to some features specific to Europe, but to its special combination of fossil fuels and access to New World resources. He shows that the conditions in early eighteenth-century Europe did not differ very much from those of China at that time, either in terms of demography, ecology, technology, or wealth. Although he shares the general world-system approach of Fernand Braudel and Immanuel Wallerstein, Pomeranz joins James Blaut (1993, 2000) and André Gunder Frank (1995, 1998) in rejecting their argument that the emergence of specific institutions in early modern Europe (including a 'capitalist mode of production') made the region more conducive to economic development than, for instance, China or Japan. By and large, circumstances that specifically favoured Europe were not internal to Europe, he argues, but the result of external global conjunctures, such as its access to silver and plantations in the New World and slaves from Africa, and the Asian demand for silver. Already in the 1780s, for instance, Britain indirectly saved about 650,000 acres per year by using New World silver to pay for its Baltic timber imports (p. 275). The list of similar calculations could be extended considerably (see Chapter 5 in this book). Suffice it to say that they add up to the conclusion that European landscapes – whether devoted to crops, pastures, forestry, mining, or urban-industrial infrastructure – for centuries have been organically connected with landscapes (plantations, clear cuts, mines, etc.) on other continents. The global environmental impacts of these interconnected and asymmetric material flows, geared to capital accumulation in core areas, are today being assessed in terms of 'ecological footprints'.

To the extent that specifically European institutions contributed to this development, Pomeranz observes, they had little to do with free markets, but rather the opposite: e.g. state-licensed monopolies, armed trade and colonization, and the projection of interstate rivalries overseas. Pomeranz clarifies his approach to world history by distinguishing between global narratives treating different regions in terms of 'comparisons' and 'connections' (p. 25). Using this distinction, we may conclude that mainstream environmental history is primarily interested in comparison, whereas Pomeranz – following the Braudelian tradition – elaborates the 'reciprocal influences' and interactions between different regions (ibid.).

In *The Unending Frontier* (2003), John Richards also highlights the early modern connections between an expanding Europe and other continents. In fourteen detailed case studies of distinct types of anthropogenic environmental change in different parts of the world, he shows how a combination of population growth and the integration of the world economy in the sixteenth to eighteenth centuries generated recurrent problems in the form of intensified land use, biological invasions, depletion of wildlife, and energy scarcity. In the Introduction, Richards aspires to 'a holistic global perspective', but modestly concedes that the selection of case studies 'may seem arbitrary, almost random' and that they are 'difficult to fit into a meaningful interpretation of environmental history and world history' (pp. 3–4). After posing questions such as, 'What is the global pattern?' and 'What shared historical processes do these case studies demonstrate?', his answer is simply to enumerate the four main kinds of environmental change mentioned above. Although he concludes that the consequences of European expansion for pre-modern ecosystems and societies were generally 'tragic', Richards finally seems compelled to condone the inevitable progress of 'brutality', 'waste', and 'degradation' toward a more 'productive', 'attractive', and 'aesthetically appealing' landscape in the newly conquered territories (pp. 10–12). Of more specific interest in his account is the decisive role he attributes to the agency of modern states in these historical processes (pp. 617–618). To once again apply Pomeranz's distinction, Richards' world history seems more concerned with global comparisons than with global connections. His position is ambiguous, however. His monumental book certainly traces the early modern emergence of a world system (a concept actually referred to on p. 17), but it does not explicitly address the fact that landscape changes within Europe in this period were recursively connected with environmental changes in other parts of the world.

The same might be said of Clive Ponting's pioneering book *A Green History of the World* (1991). Beginning with some classic 'lessons from Easter Island' (cf. also Diamond 2005), Ponting chronologically traces a cursory but global environmental history from Paleolithic foraging through Neolithic agriculture and pastoralism, population growth, European expansion, colonialism, the Industrial Revolution, urbanization, and mass consumption to global warming. Although Pomeranz's and Richards' volumes are much better documented and their arguments more sophisticated, it is striking how their narratives of energy use, European expansion, and 'the world hunt' (Richards 2003) echo Ponting's account a decade earlier. But Ponting's objective goes beyond the meticulous reconstruction of human history by conveying an explicit warning about recurrent Malthusian limits and the threat of socio-ecological collapse. Like Radkau, McNeill, Richards, and most other environmental historians, however, he does not address the analytical task of accounting for the reciprocal connections between landscape changes on different continents within a single theoretical framework – for example, the total causal web that over time connected deforestation in northern Europe, silver mining in Latin America, sugar plantations in the Caribbean, cotton plantations in south-eastern North America, and coal mining in Britain. Although brief, only Worster's (1988) early visions of global environmental history seem to anticipate the focus on *connections* that organizes Pomeranz's world-system approach.

Environmental archaeology: Redman, Erickson

The archaeologist Charles L. Redman in *Human Impact on Ancient Environments* (1999) provides several examples of human societies that have degraded their environments to the point where these societies themselves have suffered decline. The list includes early agricultural villages in the Levant, early cities in Mesopotamia, the ancient Mediterranean area, the Anasazi and Hohokam of the American Southwest, Easter Island and other instances of environmental degradation in Polynesia, and the Maya. In contrast to the historians, Redman often uses cybernetic models to represent his theoretical framework, which inclines toward the school known among anthropologists as 'cultural ecology', an approach that seeks to explain cultural forms as adaptations to technological and environmental factors. Yet, the evidence that Redman presents points in the opposite direction. Far from accommodations to environmental constraints, cultural idiosyncrasies tend to have a logic of their own, whether Levantine villagers' penchant for plaster and goats, agricultural intensification leading to salinization in Mesopotamia and the American Southwest, Polynesians' appetite for birds and birds' eggs, the obsessive construction of megalithic statues on Easter Island, or the rigidly pastoral identity of the Greenland Norse. Two fundamental messages dominate Redman's narrative: increasing scale, complexity, and intensification make societies more environmentally destructive and vulnerable, and the archaeological record indicates that such processes of environmental degradation have occurred for millennia among pre-modern societies.

This rather pessimistic view of human–environmental relations in pre-modern times contrasts with that of Clark L. Erickson, who has conducted archaeological excavations of artificial mounds and other earthworks in the periodically inundated Llanos de Mojos of Beni, lowland Bolivia (Erickson 2006; Erickson and Balée 2006). Erickson and Balée present their research as a contribution to 'historical ecology', which has become an important complement to environmental history. While both fields emphasize interdisciplinary studies of human–environmental relations over the long term, the approach of historical ecology differs from mainstream environmental history by prioritizing the use of data from natural science rather than historical sources. Archaeological excavations and forest inventories in the Llanos de Mojos converge in indicating the degree to which this is a thoroughly cultural landscape. Erickson (2006) elaborates his understanding of how the pre-Columbian inhabitants of this part of the Bolivian Amazon had 'domesticated' their landscape through the use of burning, raised fields, causeways, canals, mounds, anthropogenic islands, ring ditches, ponds, dams, and fish weirs. By regulating vegetation and water levels and facilitating transports, they had organized a highly productive socio-ecological system that sustained impressive population densities. The conclusion is that human activity may well increase biodiversity, and that there are ancient technologies for sustainably intensifying agricultural production in the Neotropical lowlands. While his explicit rejection of notions of adaptation and environmental determinism is thoroughly justified, Erickson's slightly exaggerated point that indigenous pre-Columbian populations are 'responsible for what we now call nature in the Neotropics' (p. 264) ought to be complemented with a critical account of the non-indigenous, capitalist socio-economic forces currently devastating Neotropical biodiversity. Historical ecology can demonstrate the technical feasibility of sustainable resource management in the Neotropics, but it does not provide much hope for transcending the economic system that for centuries has systematically dismantled such practices.

The contrast between Redman's and Erickson's perspectives should be clear: archaeology can be used to argue either that pre-modern societies were as ecologically destructive as modern ones, or that modern societies have something to learn from pre-modern ones regarding sustainable use of the natural environment. But neither Redman nor Erickson relate local processes of landscape change – whether in the direction of decreasing or increasing sustainability – to long-distance trade connections and incentives for export production. It is quite conceivable, however, that much of the stimulus for ancient intensification of production – and concomitant environmental change – in both the Old and the New World derived from such distant sources of demand (cf. Chapter 7). Whether extractive or agricultural intensification occurred to provide directly for distant consumers, or for an expanding local population engaged in export production, the global context should be an essential component of accounts of environmental transformations.

Environmental geography: Diamond, Simmons, Moore

Jared Diamond's book on *Collapse* (2005) has been widely read and cited. This is probably not only because of his clear and accessible prose – his previous book *Guns, Germs, and Steel* (1997) earned him a Pulitzer Prize – but because its central theme, the ecological self-destruction of past and present societies, appeals to a widespread concern with the sustainability of contemporary industrial civilization. Diamond's main ambition is to scrutinize a selection of archaeological and historical examples of societal 'failure' and 'success' in order to draw general conclusions on how modern societies ought to behave so as to increase their chances of survival. He summarizes and popularizes relevant archaeological research on socio-environmental collapse on Easter Island, Pitcairn and Henderson Islands, the Anasazi, the Maya, and Norse Greenland, contrasting such 'failures' with the historical 'success' of Iceland, the New Guinea highlands, Tikopia, and Japan. The book begins with a long chapter contemplating the problems and prospects of modern Montana – Diamond's cherished summer resort – and devotes four chapters to a motley assortment of other modern case studies, including Rwanda, the Dominican Republic, Haiti, China, and Australia. The penultimate chapter assesses the contrasting environmental records of some industrial corporations, including two oil companies, and the final chapter includes a section on the author's experience of living in Los Angeles.

This remarkable mix of case studies predictably fills the book with peculiar analogies. The mere idea of juxtaposing the fate of a few dozen prehistoric residents of Henderson Island with the prospects of modern nations such as China or Australia – as comparable 'societies' confronted with similar dilemmas – seems a bizarre confusion of scales. The list of explicit analogies includes comparisons between prehistoric Easter Islanders, on one hand, and modern Montana farmers (p. 75), Hollywood moguls (p. 98), Romanians (p. 110), and Rwandans or Haitians (p. 151), on the other; between Easter Island and 'the whole modern world' (p. 119); between Mangareva Island and the United States (p. 120); between Chaco Canyon Anasazi and citizens of Rome and London (p. 150) or New York (p. 154); between Maya kings and modern American CEOs (p. 177); between the Greenland Norse and 'oil-importing Americans' (p. 267), the Soviet Union (p. 272), rioters in Los Angeles (p. 273), and the Bush administration (p. 425); between Japanese shoguns and president Kennedy (p. 439); and so on. This proliferation of startling analogies is ironic, since Diamond himself lists 'false analogy' as one reason why a society may fail to anticipate problems (p. 423).

Jared Diamond is a biogeographer whose PhD thesis was devoted to the ornithology of New Guinea. As so often when a scholar with a background in natural science turns to human history, there is a disturbing silence on the role of specificities of culture and social structure in accounting for historical processes and events. Diamond's (1997) geographical determinism is seductive in its simplicity (but cf. Blaut 2000 for a trenchant critique) and appears to reflect a desire for simple, material explanations of world history that extends beyond biogeography into much mainstream historiography as well (cf. Crosby 1972, 1986). The fifteen-page index in *Collapse* does not even include an entry on 'culture' or 'cultural'. Diamond's assumptions about failures in societal 'decision-making' (p. 420) are also politically naïve in underestimating the role of power structures and irreconcilable conflicts of interest throughout human history. Ultimately, it is his notion of 'societies' as a unit of analysis that is misguided. Neither the Maya, the Anasazi, medieval Iceland or Japan were self-contained managerial units that could 'choose to fail or succeed' – a rhetoric more properly evoking a US presidential administration or the board of an oil company. All these populations, not to mention those of modern Montana, Rwanda, and Haiti, should be recognized as components within larger regional or global systems of societal reproduction, within which some subsystems progress and accumulate at the expense of others. When Diamond expresses hopes that, by learning from the past, 'we may keep on succeeding' (p. 3), it is not evident who is to be included in the category 'we'.

Paradoxically, Diamond's recurrent recognition of the significance of long-distance trade, interdependency, and globalization does not prompt him to abandon his atomistic approach to 'societies' as geographically delineated populations managing their own destiny. The concept of a world system seems as alien to Diamond as a serious penetration of non-European cosmologies or social structures. In order to understand the specific trajectories of human societies over the past millennia, it seems, all we need is a well-read ornithologist with an interest in archaeology and climate change. Social science theory is completely absent, and the only reference to anthropology is a scornful mention of its generally sceptical attitude to reports of cannibalism (p. 152), which Diamond does not hesitate to identify in most of the cases he lists as 'failures'.

Another paradox is that Diamond's refreshing – if naïve – criticism of self-serving power elites inflicting harm on others through their 'bad' behaviour (pp. 427–431) in no way seems to shake his confidence in industrial capitalism and the imperative of making profits (pp. 441–442). Although he makes a point of showing how environmental and political problems tend to go hand in hand, and although he often demonstrates an awareness of how the wealthy and powerful are the last to suffer from environmental deterioration – for instance, by being able to import resources and export garbage (p. 370) – he seems unaware of the burgeoning literature on 'political ecology', 'environmental load displacement', and 'environmental justice'. His recipes for sustainability thus have little to offer beyond the familiar invocations of consumer power (pp. 484–485), new values (pp. 432–433), First World restraints (pp. 496, 519), better decision-making routines, and courageous leadership (p. 440).

Although a geographer with a quite different background, Ian G. Simmons in *Environmental History: A Concise Introduction* (1993b) offers only patchy references to the global context of landscape transformations. He briefly mentions instances of the displacement of environmental loads from Britain (pp. 126, 128, 135, 145) and Japan (pp. 154, 155), but again there is no theoretical framework for systematically treating such drivers of landscape change. His 'concise' introduction to environmental history meanders in a bewildering variety of directions, from the 'Pleistocene overkill', salinization, and overgrazing to the environmental perspectives of Taoism, gardening, and American wilderness conservation. As we have seen, this tendency toward amorphous narratives is a recurrent feature in several of the major contributions to environmental history. Simmons' recent volume *Global Environmental History 10,000 BC to AD 2000* (2008) fares no better in terms of acknowledging theoretical frameworks from social science. Again, the concept of 'global' here seems to refer to the geographical range of examples rather than to a coherent framework for understanding their interconnections.

Against this background, it is encouraging to find the geographer Jason W. Moore (2000, 2003, 2007, 2009) exploring the potential of a Braudelian world-system framework for understanding global environmental history. To those of us who remain impressed by Braudel's (1992 [1979]) pioneering insights, Moore's studies are welcome, for they constitute a series of long footnotes to the Braudelian edifice. In fact, in terms of theoretical framework, Moore's own quotes from Braudel say it all – no theoretical advance is required in order to account for the empirical data that Moore assembles:

> Europe, because of her very expansion, was acting as if she had decided to delegate the trouble of handling of the mining and metallurgy industries to dependent regions on her periphery. In the heart of Europe, not only were falling yields limiting profits, but the 'fiery furnaces' were destroying forestland, and the price of wood and [charcoal] was becoming prohibitive, so that the blast furnaces could only operate part of the time, thus immobilizing fixed capital to no purpose. Meanwhile wages were going up. Small wonder then that the European economy as a whole applied to Sweden for iron and copper; to Norway for copper; before long to distant Russia for iron; to America for gold and silver…
>
> (Fernand Braudel 1992 [1979], vol. 2, p. 325; quoted in Moore 2009:54)

Here is a framework for understanding how, for instance, in the sixteenth century, the environmental devastation associated with silver mining in Bolivia was intertwined with Baltic logging financed by the Dutch. Such insights should be central to a truly global environmental history. To be sure, Moore strays no further from the Braudelian framework than to invite the same charges of Euroexceptionalism as have been directed at his paragon by Frank (1995) and Pomeranz (2000), but his determination to interweave the landscape trajectories of different continents is commendable and should inspire future research on global environmental history. On closer scrutiny, Moore's accounts fail to convincingly identify a decisive discontinuity or 'rupture' between pre-capitalist and capitalist modes of production, e.g. in terms of the metabolic rift between countryside and city (2009:5–6), the importance of urban grain imports (p. 18), or the pace of environmental transformation (p. 19), but this is a long-standing discussion that will no doubt continue to haunt Braudelian historiography until more stringent analytical approaches have been consolidated (cf. Hornborg 2000; Hornborg *et al.* 2007). In the meantime, it is gratifying to note that Moore belongs to an increasing number of scholars applying world-system perspectives to environmental history (Goldfrank *et al.* 1999; Bunker and Ciccantell 2005; Jorgenson and Kick 2006; Hornborg *et al.* 2007).

Perspectives on 'collapse'

The concern with socio-ecological collapse, which animates Jared Diamond's (2005) account, is a recurrent feature of studies in environmental history. Much of the discussion has focused on the fate of polities that have left conspicuous remains in the form of impressive monuments, e.g. ancient Mesopotamia, Rome, Maya, and Easter Island. Although distorted by Diamond's moralizing discourse on 'how societies choose to fail or succeed', the concept of 'collapse' has been well defined by anthropologists and archaeologists (Tainter 1988; Yoffee and Cowgill 1988) as a sudden loss of societal (political or economic) integration and complexity, resulting in a fragmentation into less inclusive and more autonomous social units. Several studies suggest that disturbances in long-distance trade relations may have contributed to the decline of ancient polities in Mesopotamia (Yoffee 1988) and tenth-century Mexico (Rathje 1973; Webb 1973; Jones 1979). Others have emphasized how growing sociopolitical investments and intensification of resource use may have precipitated ecological collapse in Yucatán (Culbert 1973, 1988; Renfrew 1978) and on Easter Island. In *The Collapse of Complex Societies*, the anthropologist Joseph Tainter (1988) persuasively argues that societies collapse when they suffer from declining returns on 'investments in complexity'. This account has the advantage of being applicable to the wide range of past societies where overinvestment in some particular kind of infrastructure (e.g. Roman armies, Mayan temples, Easter Island statues) is conventionally viewed as a factor contributing to their decline. In suggesting that specific cultural strategies for problem-solving can recursively aggravate the problems – whether, for example, military, political, ideological, or ecological challenges – and generate vicious downhill circles, Tainter's explanation is intellectually appealing. Moreover, it prompts us to scrutinize escalating investments in technological infrastructure (e.g. for transports, environmental protection, medical care, social rehabilitation, etc.) in modern societies.

In contrast with superficial accounts such as Diamond's *Collapse* or Sing Chew's (2001, 2006) books on 'Dark Ages' (for critiques see Moore 2001; Tainter 2006; Widgren 2007b), Tainter also dedramatizes processes of societal decline by observing that collapses or 'Dark Ages' have rarely been as calamitous as historiographers tend to imagine. Socio-political collapse does not necessarily imply ecological degradation or major population loss. Although marred by the failure to distinguish between past socio-political complexity and current ethno-linguistic identity (cf. McNeill 2010), this conclusion is underscored by Patricia McAnany and Norman Yoffee (2010) in their edited collection *Questioning Collapse*. Regardless of the contested issue of the scope of human suffering, the recurrent phenomenon of societal decline deserves theoretically and empirically well-informed elucidation. Tainter's formalist model needs to be complemented with close, case-by-case attention to the cultural specificity of fetishized, counterproductive strategies for solving problems of sustainability. In addition to the mathematical abstraction of curves depicting the marginal productivity of different kinds of societal investments, we need to understand how idiosyncratic and unpredictable cultural conceptions about the nature of the world can represent specific material infrastructures (such as temples or monolithic statues) as technologies that are indispensable for human survival. Technological systems rely on *beliefs* about their efficacy and efficiency. From an abstract, comparative perspective, they can be defined as culturally accepted strategies for achieving culturally accepted goals. As suggested in previous chapters, such insights could help us expose those specific varieties of fetishism that may be pushing modern capitalist society itself toward a downhill slope (Hornborg 2001b).

This is not to suggest an abandonment of formalist models, however. Beyond the cultural specificities we can certainly discern abstract regularities over the millennia, as processes of capital accumulation, unequal exchange, and environmental load displacement generate recurrent socio-ecological polarizations that lead to crisis, contradiction, and decline. Given the requisite analytical precision and empirical care, Tainter's model can usefully be integrated with approaches rooted in world-system analysis. Societal collapse can be a consequence of increasing isolation from as well as integration into more inclusive systems of exchange. The former trajectory can be exemplified by cases as diverse as the medieval Norse colony on Greenland, the Classic lowland Maya, and ancient Assyria, the latter by innumerable indigenous populations (including that of Easter Island) upon contact with the colonial world system. In comparing such cases, however, it is important to avoid juxtaposing 'societies' at grossly incommensurable scales, such as two medieval villages on Greenland versus the entire Chinese empire (cf. Diamond 2005; McAnany and Yoffee 2010). The comparative study of socio-political collapse thus needs to consider what a 'society' is, how the units of analysis are bounded, and who the social actors are that 'fail' or 'succeed'. It also needs to scrutinize the notion of 'success' itself, as illustrated by the case of the contemporary Netherlands, whose affluence to a large extent hinges on environmental load displacement to Indonesia, Brazil, and West Africa (McNeill 2010:360).

Accumulation, 'value', and ecologically unequal exchange: toward an analytical framework for global environmental history

Central to a world-system perspective is the concern with accumulation, i.e. the strategies of different groups to enrich themselves through various kinds of exchange. The concept of accumulation (as opposed to the notion of 'growth') is generally taken to presuppose some kind of unequal exchange, and unequal exchange in turn tends to be defined in terms of asymmetric transfers of some kind of 'value'. For Karl Marx and his most orthodox followers, the notion of 'surplus value' accumulated by capitalists is thus founded on a theory of labour value, while for the earliest proponents of a theory of *ecologically* unequal exchange (Bunker 1985, 2007; Odum 1988; Odum and Arding 1991), accumulation is tantamount to a net transfer of energy or 'natural values'.

As opposed to such conceptual dependencies of theories of 'accumulation' and 'unequal exchange' on the notion of 'value', I believe that it is imperative to maintain an analytical distinction between the material/biophysical and the cultural/semiotic dimensions of exchange. It is very obvious that the 'value' or attractiveness of a commodity for a given consumer hinges on the cultural preferences of that consumer (Baudrillard 1972; Sahlins 1976; Bourdieu 1984), rather than on the investments of labour or energy made in its production, and that the former cannot be reduced to the latter. Any history of consumption will make it abundantly clear that the first condition for accumulation is that there is a cultural demand for the commodity in question (e.g. Wolf 1982; Pomeranz and Topik 1999). In this respect, economists of all persuasions should be in agreement. But, contrary to mainstream economists, we must recognize that a second condition for accumulation is the material organization of production. It is this biophysical dimension of economic processes that the mainstream economists' preoccupation with 'utility' neglects, and that has been the common denominator of the many materialist challenges to this preoccupation from Karl Marx to ecological economics. A crucial task is to offer such a challenge, which acknowledges the biophysical dimension but without equating it with 'value'.[2]

Rather than posit an 'unequal exchange of value', accounts of accumulation need to combine, on one hand, an understanding of the requisite cultural constructions of consumption and market demand, and on the other, an analysis of the specific organization of material resources employed to cater to those demands. This combination of qualitative/semiotic and quantitative/biophysical knowledge is rarely fostered in the current division of labour between disciplines, yet it is essential for understanding the relations between culture, economy, and ecology. Accumulation, and by extension global environmental history, hinges precisely on these relations.

For several millennia, trade has been driven by the aspirations of various groups to enrich themselves, i.e. to accumulate. Long-distance traders saw opportunities to profit from geographical discrepancies between different cultural valuations of commodities. Local elites were enthusiastic consumers of exotic imports that helped to communicate their privileged social positions. In the regions of the world system where these imports originated, local producers were encouraged or coerced to invest labour and transform landscapes in response to such distant demand. This general understanding of world economic history is applicable to a very long list of traded commodities over the past few millennia, including a wide variety of exotic foodstuffs, spices, drugs, animal parts, textiles, dyes, metals, and manufactures such as porcelain (cf. Pomeranz and Topik 1999). The social and environmental impacts of such export production, particularly over the last five centuries, have no doubt affected most of the land surface of the Earth. We need only think of the vast impacts of the trade in cotton, silk, sugar, coffee, tea, tobacco, silver, and furs.

In order to analytically clarify the changing environmental dimensions of the history of world trade, it is useful to distinguish between two main strategies for enhancing accumulation through investments in capital, widely conceived. 'Capital' is here defined as investments of labour time and natural resources for the purpose of increasing the productivity of land or labour. The universal incentive for capital accumulation is to increase output, but the extent to which such increases are relative to inputs of labour time or of landscape space depends on circumstances. The rationality of different strategies for capital accumulation, in other words, is socially constructed. With the integration of increasingly wider global markets, the ambition to increase productivity has generally been connected to the imperative to increase competitiveness, which usually means producing commodities at a relatively low cost per unit produced. Measures to lower costs may include increasing efficiency of production (e.g. through mechanization) as well as minimizing costs for labour, land, energy, and raw materials. To increase efficiency generally means to increase the volume of production, in order to benefit from so-called economies of scale.

Capital accumulation for the purpose of increasing the productivity of land has been referred to by Harold Brookfield as 'landesque' capital (cf. Widgren 2007a). This category includes inalienable modifications of land such as irrigation or drainage canals, terraces, raised fields, forest clearance, stone clearance, and soil improvement. In non-industrial societies throughout history, all such changes of the land have required considerable inputs of human labour. Their rationale has universally been to increase output per unit of land, even if it should imply increasing inputs of labour. The accumulation of landesque capital has often been recursively connected to the concentration of human populations in larger communities with more complex division of labour, including processes of urbanization. The increased agricultural output per unit of land has made such demographic concentration and socio-economic complexity possible. Conversely, in representing a valuable resource coveted by militant neighbours, the investment in landesque capital has required access to larger populations for purposes of defence. Larger populations have in turn demanded more socio-political complexity and agricultural output, and so on. Even if, as Widgren (ibid.) points out, landesque capital has often permanently improved the conditions for sustainable human land use, there are also examples of adverse effects. Among the diverse environmental impacts of various forms of landesque capital are salinization, deforestation, drainage of wetlands, depletion of groundwater, eutrophication, carbon dioxide emissions, and erosion following abandonment.

If landesque capital is defined as non-detachable investments in land for the purpose of increasing its productivity, what we usually think of as 'capital' (pertinently referred to by Amartya Sen as 'labouresque' capital) should be defined as investments for the purpose of increasing the productivity of labour. This category of investments can be subdivided into two analytically distinct but interrelated types: (1) education and training resulting in specific types of competence and skill, and (2) technology, widely defined. Beyond the sophisticated local efficiencies of pre-industrial technologies and 'traditional ecological knowledge', the accumulation of labouresque capital has generally been recursively connected to a successful engagement in trade, measured as a net appropriation of biophysical resources such as energy, embodied land, or embodied labour. A continuous net gain in access to such resources can be converted into technological growth, as illustrated by the contemporary imports of fossil fuels to the United States, or by nineteenth-century imports of cotton fibre to England (Chapter 5). If the accumulation of landesque capital has been recursively connected to population growth, then, the accumulation of labouresque capital is recursively connected to unequal exchange. It would be superfluous to exemplify the environmental impacts of technological intensification, whether in the vicinity of industrial factories, the distant sources of their raw materials, the disposal of garbage, or the atmosphere.

This is not the place to rewrite the global history of human–environmental relations in terms of capital accumulation and unequal exchange. Suffice it to say that such a project would be both feasible and essential. Once we rid ourselves of the ambition to ground our understanding of unequal exchange in some putatively objective notion of 'value' (Chapter 6), we can focus on the objectively quantifiable net transfers of energy, embodied land, and embodied labour in world trade. Such material transfers have historically been geared to production processes catering to the most diverse cultural desires, whether porcelain from China, cotton textiles from Gujarat (or British imitations thereof), hats from Canadian beaver, ornaments from African ivory, or the taste of Moluccan nutmeg, Mexican cacao, or Virginian tobacco. Global histories of cultural desire are continuously being written, but so far no systematic global history of the environmental impacts of these desires, or of the production processes organized to cater to them. The cultural attribution of 'value' to commodities such as sable, silver, cinnamon, coffee, or Coca-Cola should not be analytically confused with the biophysical changes in ecosystems subjected to their production. This of course applies no less to modern industrial exports such as cars, cellular phones, and computers. A truly global environmental history would need to systematically examine (1) how particular constellations of cultural demand have encouraged specific strategies of accumulation and export production, (2) how such interconnected strategies of accumulation have entailed net transfers of energy, embodied land, and/or embodied labour, and (3) how these processes of extraction, production, and transport have affected societies and environments in different parts of the world system. It would also need to distinguish between environmental problems deriving, respectively, from biophysical impoverishment versus biophysical overload. While extractive zones will tend to experience loss of biodiversity, topsoil, fish stocks, and other vital assets, world-system centres have historically suffered from smog, acidification, eutrophication, accumulation of heavy metals, and problems with the disposal of solid waste. Whereas the former problems result from removal of resources, the latter are associated with a concentration of the use of matter and energy. The emission of carbon dioxide from combustion of fossil fuels is a tangible illustration of the global displacement of entropy associated with capital accumulation. The issue of 'climate justice' (Roberts and Parks 2007, 2009) is founded on the fact that such emissions are largely the result of technological accumulation and energy use in the North, whereas their deleterious consequences disproportionally afflict the South. Using the atmosphere as a sink for carbon entropy, in other words, is yet another example of environmental load displacement. Such environmental inequalities, of course, recur at various levels of scale within the North as well as the South, but a truly global environmental history must acknowledge that the metabolic rift (Foster 1999; Foster *et al.* 2010) ultimately polarizes populations and landscapes at the planetary level.

Over the long term, diverse combinations of 'landesque' and 'labouresque' capital have been accumulated to cater to the demands of markets at varying distances. At our current point in historical time, of course, it is particularly important to scrutinize the role of fossil fuels in the past two centuries of capital accumulation and environmental change, but the general pattern has precedents going back several millennia. The large-scale environmental impacts of the earliest urban civilizations in Mesopotamia, India, and China were generally restricted to deforestation and salinization resulting from agricultural production and the demand for timber, firewood, and charcoal. Timber was required for urban construction as well as shipbuilding, while firewood and charcoal were used for heating, brick-making, metallurgy, and various other kinds of proto-industrial manufacture. Although not as geographically extended as in more recent societies, the polarization of centres of accumulation and impoverished extractive zones was evident even in these early civilizations. Such processes became more pronounced with the Roman Empire, where the distant appropriation of energy in the form of slaves and grain generated social and ecological impoverishment over much of the Middle East, North Africa, and Europe. Landscapes in many areas of the Old World were successively transformed by export production of goods such as wheat, wine, olive oil, wool, cotton, rice, silk, and spices. In the Middle Ages, the continued long-distance exchange of manufactures and preciosities across the extent of the Old World stimulated, for instance, the industrial manufacture of porcelain and silk in China, cotton textiles in India, and glass and woollens in northern Italy, as well as the extraction of ivory and gold in Africa and amber and furs along the Baltic. All such activities had ecological repercussions, whether intentionally (e.g. spices, silk, cotton, sheep) or not (e.g. ivory, charcoal, overhunting). Growing human populations also exerted an increasing pressure on available agricultural land and wood fuel. By the fifteenth century, several areas of the Old World were experiencing serious ecological constraints. The geographical position of Europe in the sixteenth century gave it access to New World silver, furs, forests, and agricultural land, which over the next five centuries stimulated new and expansive strategies for capital accumulation through long-distance trade, colonialism, slavery, and industrialization. The British shift to fossil energy for industrial production decisively changed the conditions for accumulation, but was a logical fusion of older industrial technologies in metallurgy (coal combustion) and water-powered textile production (the mechanical conversion of linear to rotating movement). The use of fossil energy in mechanization importantly decoupled energy use from land constraints, encouraged a vastly expanded trade in bulk goods (including food) across the world, and enabled European industry to oust its competitors (including the Indian textile industry) on the global market.

This very brief synopsis of five millennia of capital accumulation is neither new nor particularly controversial, but the important point is that a world-system perspective should prompt us to view the various environmental changes precipitated by these processes as systematically interrelated. For example, deforestation and environmental transformations associated with landesque capital have for millennia in various parts of the world been inextricably linked to distant markets for timber, charcoal, and agricultural products, which in turn have been linked to urbanization, marine trade, industry, and various kinds of consumption. The ecological impoverishment of North Africa and much of the Mediterranean area was directly related to the accumulation of infrastructure, armies, and slave labour in urban Rome. The environmental changes associated with sheep, cotton plantations, and silkworms have been causally connected to textile industries in Europe, India, and China, and to their distant markets. The European and Asian markets for elephant ivory transformed the ecology of East Africa (Håkansson 2004, 2007). Since the economic integration of the Old and New Worlds in the sixteenth century, complex webs of connections have intertwined the environmental consequences of silver mining in Bolivia, sugar plantations in the Caribbean, and the trade in North American furs, Baltic timber, African slaves, Indonesian spices, and Chinese tea.

If a world-system perspective is crucial for understanding the local details of environmental history, ecology is thus no less fundamental for understanding world-system history. To trace the metabolic flows of world systems requires a basic familiarity with their biophysical aspects, such as the use of energy and land, the displacement of entropy (including carbon dioxide emissions), and ecologically unequal exchange (Hornborg 2001b; Frank 2007; Jorgenson and Clark 2009). It also requires a radical rethinking of our everyday conception of technology as an accumulation of material infrastructure, or capital, that should be understood more as the product of a global zero-sum game than as a cornucopia.

5 The unequal exchange of time and space

I have long been pursuing an understanding of industrial capitalism as to some extent a zero-sum game. In this work I have been using the Second Law of Thermodynamics (the so-called Entropy Law) as a point of departure, arguing that the build-up and maintenance of industrial infrastructure necessarily requires a continuous net input of free or available energy (technically referred to as *exergy*, and closely related to the concept of *negative entropy*; cf. Schrödinger 1944; Georgescu-Roegen 1971). Purely analytically, it is possible to conclude that industrial infrastructure – whether a factory, an industrial city, or the global technomass – must maintain an unequal exchange of free energy with its hinterland in order to survive and grow. It is also possible to analytically conclude that, under market conditions, this unequal exchange will be orchestrated by the terms of trade between industrial and extractive sectors of the national and global economy, i.e. the rate at which industrial manufactures are exchanged for fuels and raw materials. The concept of unequal exchange in this sense is objectively specified rather than normative, and can be applied to several other possible metrics of trade, including material flows (Fischer-Kowalski 1998). Georgescu-Roegen argued that not only energy but materials, too, suffer irreversible dissipation in economic processes, and that the concept of entropy applied to such processes should be understood in terms of the generation of increasing energetic and material disorder as a by-product of the local creation of cultural and technological order or structure. Prigogine's concept of *dissipative structures* has been applied to the asymmetric flows of entropy and negative entropy between different sectors of global society (Hornborg 1992; Clark 1997). Combined with a world-systems perspective (Frank 1966; Wallerstein 1974–1989), such an understanding of industrial capitalism provides a theoretical framework for interpreting the socio-ecological logic by which environmental problems tend to be unequally distributed between different sectors of the global population.

Owing largely to its foundation in thermodynamics, this analytical work has been difficult to translate into quantitative, empirical research. Actually traded quantities of exergy and negative entropy, as embodied in commodities, are inherently difficult to estimate. However, it is possible to detect such structural asymmetries in trade by converting statistics on commodity flows into quantities of embodied land and embodied labour. Both these factors of production can be sources of exergy for the accumulation of capital, but have the advantage of being quantifiable, e.g. in annual hectare yields and in hours of human labour. Whether or not the traded commodities are actual physical sources of exergy for capital (e.g. fuels, labour, food), it should be illuminating to calculate net transfers of embodied labour and land in relation to exchange values. The simpler a system of trade relations is, the easier it should be to make such calculations. I have thus chosen to investigate, from this perspective, certain aspects of British overseas trade in the mid-nineteenth century. The years around 1850 are particularly interesting in that it is possible to obtain a wide range of fairly reliable data for these years, during what is a comparatively early and expansive period in British industrialization. These data have the advantage not only of being fairly simple and easy to synthesize, in comparison with modern trade, but also of illuminating the very origins of industrial capitalism. It is not unreasonable to propose that global patterns of asymmetric resource exchange established in the early phases of industrialization have continued to reinforce themselves through the continuing technological superiority (i.e. the accumulation of capital) that they engendered in north-western Europe. Recent statistics supporting such an interpretation include estimates of so-called ecological footprints of different nations (Wackernagel and Rees 1996).[1]

In approaching the statistics on nineteenth-century British trade, I initially posed the following questions:

1 How many annual hectare yields and hours of human labour were embodied in the British imports of raw cotton, wheat, and other commodities?
2 How many annual hectare yields and hours of human labour were embodied in the British production of textiles, and at which rates were they exchanged for those embodied in imports?
3 How many domestic (British) annual hectare yields and hours of human labour were liberated for other purposes by the displacement of land and labour inputs to other nations?

The answers to these questions are not as readily available in the standard literature on British economic history as I had anticipated, which suggests that such questions have been far from prominent, to say the least, in the voluminous research on the Industrial Revolution. In fact, my impression is that the conventional economic discourse on industrialization conspires to keep such questions – and their answers – out of view. The figures that I have finally arrived at, often through convoluted deductions from statistics assembled to illuminate very different issues, are presented in Table 5.1. For each calculation I have provided a note explaining how the figures were calculated. Some of these figures may seem uncertain, but in this first approximation I consider the questions posed and methodology used more important than the minute details of mathematical accuracy. I would welcome suggestions on how to improve accuracy in these estimates, whether by consulting other sources or applying other methods of calculation. The results arrived at, however, are probably not too far off the mark to lack significance for a rethinking of industrialization (capital accumulation) in terms of the appropriation of land and labour from elsewhere.

From these data, a number of observations can be drawn about the global economies of (natural) space and (human) time that were recursively linked with technological development, or capital accumulation. To begin with, we can now provide provisional answers to the three questions raised at the outset:

1 The 1850 import of 223,623 tons of raw cotton from the American South represented over 616 million hours of (mostly slave) labour and the annual yield of over 1.1 million hectares of agricultural land. If projected onto total imports (254,921 tons), the flow of raw cotton to England in 1850 embodied over 702 million hours of labour and the yield of 1.26 million hectares. If British figures on wool production are generalized to apply also to imported raw wool, wool imports in 1850 (27,170 tons) represented over 26.8 million hours of labour and 0.75 million hectares. In the same year, imports of wheat represented over 985 million hours and the productivity of 2.95 million hectares abroad. Imports of these three commodities alone – cotton, wool, and wheat – in 1850 represented 1,714 million hours of labour and a year's harvest from almost five million hectares of land.
2 The 1850 production of cotton manufactures in Britain (226,879 tons) represented around 272 million hours of British labour. The production of woollen manufactures in the same year (66,303 tons) reflects a total of over 549 million hours of British labour[2] and the productivity of 2.44 million British hectares of pasture and hay.[3] In order to be able to examine relative exchange rates of embodied labour and land in exports and imports, I have calculated the purchasing power of £1,000 around 1850 in terms of raw cotton, cotton manufactures, raw wool, woollen manufactures, domestic wheat, imported wheat, and guano (Table 5.2).

Table 5.1 Estimates of inputs of land and labour in some key commodities in British overseas trade around 1850. Measures are in hectares, hours, metric tons, and British currency *c*.1850

Cotton	*Years*	*Quantity*	*Wool*	*Years*	*Quantity*	*Wheat*	*Years*	*Quantity*
Hours of labour required for production of one ton of raw cotton in American South[1]	1850	2,755 h	Hours of labour required for production of one ton of raw wool in England[2]	1850	988 h	Hours of labour required for production of one ton of wheat imported to England[3]	1850	1,144 h
Hectares of land required for production of one ton of raw cotton in American South[4]	1850	4.95 ha	Hectares of land required for production of one ton of raw wool in England[5]	1867	27.7 ha	Hectares of land required for production of one ton of wheat imported to England[6]	1850	3.43 ha
Tons of raw cotton imported from American South to England[7]	1850	223,623 t	Tons of raw wool produced in England[8]	1850–1854	61,236 t	Hectares of land required for production of one ton of wheat in England[9]	1851	1.14 ha
Tons of raw cotton imported to England (total)[10]	1850	254,921 t	Tons of raw wool imported to England[11]	1850	27,170 t	Tons of wheat imported to England[12]	1850	861,790 t

Hours of labour required for production of one ton of cotton manufactures in England[13]	1849–1851	1,200 h	Hours of labour required for production of one ton of woollen manufactures in England[14]	1850	6,968 h	Hours of labour required for production of one ton of wheat in England[15]	1851	394 h
Quantity of cotton yarn produced in England from one ton of raw cotton[16]	1828–1861	890 kg	Quantity of woollen manufactures produced from one ton of wool[17]	1741	750 kg	Tons of wheat harvested in Gt. Britain[18]	1849–1854	3,620,000 t
Tons of cotton manufactures produced in[19] (and exported from[20]) England	1850	226,879 t (168,910)	Tons of woollen manufactures produced in England[21]	1850	66,303 t	Tons of guano imported to England[22]	1851–1853	165,350 t
Exchange value of one ton of raw cotton[23]	1850	£84.46	Exchange value of one ton of raw wool[24]	1850	£71.88	Exchange value of one ton of guano[25]	1851–1853	£15.89
Exchange value of one ton of cotton manufactures exported from England[26]	1850	£293	Exchange value of one ton of woollen manufactures exported[27]	1854	£342	Exchange value of one ton of wheat imported to England[28]	1849–1854	£222

Notes

1 McDonald and McWhiney (1980:1098) provide the figure 0.8 hand-hours to produce a pound of cotton on an Alabama plantation in 1850. One ton = 2.204 lbs = 2.755 hours.

continued

2 The average weight of a fleece of wool (presumably an average for ewes and lambs, i.e. sheep under one year old) in 1688–1695 was estimated at 4 lbs (Deane 1957:209) or 1.81 kg, while a more conservative estimate in 1741 suggests 3.5 lbs (ibid., 211) or 1.58 kg. Modern averages from British breeds and two shearings annually suggest a wool harvest of 4.2 kg per ewe (Sjödin 1974:384). One way of estimating actual wool output is to use data on numbers of sheep and total wool clip from a year (1867) when both figures are available. Mitchell (1962:82, 84, 190) gives 28,919,000 sheep (ewes and lambs) in Great Britain and 4,836,000 in Ireland in 1867 and a domestic wool clip in the United Kingdom of 163,000,000 lbs = 73,936,800 kg. A total of 33,755,000 sheep in the United Kingdom thus yielded 4.8 lbs = 2.1 kg of wool per sheep. The total clip divided by 20,483,000 ewes gives 3.6 kg of wool per ewe and 278 ewes per ton of wool produced. Labour input per ewe has been estimated at four hours per ewe (Sjödin 1974:400) or 1,112 hours for animal care and 864 hours for fodder production and fencing. The latter figure builds on Collins' (1969:460) estimate of mowing rate with scythe (albeit for corn rather than hay) at around one acre per worker-day, plus an estimated additional day per acre for drying and harvesting = two days/acre = approximately five worker-days per hectare, and a requirement of 1.5 kg hay per ewe for 120 days × 278 ewes = fifty tons of hay. With worker-days estimated at ten hours and average hay yields at 1.4 tons per acre = 3.45 tons per hectare in 1885 (Mitchell 1962:90), this means about 724 hours in fodder production per ton wool produced. With an approximated additional two weeks (120 hours) spent annually on mending fences, this gives 864 hours, and a total of 1,976 hours for 278 ewes. Estimating that wool sales accounted for roughly 50 per cent of income from a flock, this figure has been reduced by half to 988 hours. This latter estimate is based on commodity prices for 1854–1860 provided by Schlote (1952:116), according to which wool was valued at £7 per 100 lbs = £0.154 per kg or £0.55 per ewe and year. The value of meat production per ewe and year has been estimated at around £0.7, based on a yield of 20 kg × £0.035. This estimate is based on a comparison of meat prices as given by Schlote (ibid.): Bacon = £2.6/cwt; ham = £3.2/cwt; salt beef = £2/cwt; salt pork = £2.4/cwt. Mutton has been estimated at £1.8/cwt. One hundredweight (cwt) = 50.8 kg.

3 Mitchell (1962:100) lists the principal sources and quantities (in thousands of hundredweights) of wheat imports to the United Kingdom in 1850: Russia = 2,766,000 cwt, Prussia = 3,609,000 cwt, Canada = 38,000 cwt, USA = 436,000 cwt. If 1 cwt = 50.8 kg, the metric equivalents are: Russia = 140,512 tons, Prussia = 183,337 tons, Canada = 1,930 tons, USA = 22,148 tons. Clark (1993:228) estimates the output per worker in bushels of wheat in Germany in 1850 as 113 bushels and in Russia in 1870 as eighty bushels. Using the average conversion factor from bushels to hundredweights of wheat (1.81 bushel = 1 cwt) in the period 1910–1939 (Mitchell 1962:90, 102), I have counted one bushel of wheat as equivalent to 28 kg. This gives an output per worker in Germany of 3,164 kg and in Russia of 2,240 kg. Assuming 3,000 hours per work-year, this would mean 949 hours per ton of wheat in Germany and 1,339 hours per ton in Russia. Considering that British wheat imports from Prussia and Russia are of similar proportions, whereas imports from North America are still marginal in 1850, I have entered the average of Prussian and Russian conditions = 1,144 hours per ton.

4 According to McDonald and McWhiney (1980:1096), the average yield per acre for the American South as a whole in 1850 was about 180 pounds of ginned lint. 180 lbs/acre = 201.67 kg/ha = 4.95 ha/ton.

5 Modern agricultural handbooks estimate 6.2 ewes per hectare pasture and hay (Sjödin 1974:384). I have reduced this figure to five ewes per hectare in the mid-nineteenth century. One way to calculate the number of hectares required per ton of wool would be to simply divide 278 ewes by five, which gives 55.6 hectares. If we divide the total number of ewes in the United Kingdom in 1867 (20.483.000) by five, we get 4.096.600 as an approximate number of hectares (pasture and hay) allocated to sheep. As the total clip in 1867 was 73.936.800 kg, the amount of wool produced per hectare can then be estimated at 18 kg, which again gives 55.6 hectares per ton of wool. Estimating the value of wool as about 50 per cent of total output of sheep in monetary terms, the number of hectares should be reduced by half to 27.7.

6 Clark (1993:228) estimates the output of wheat per acre in Germany in 1850 at 7.1 bushels and in Russia in 1870 at 3.0 bushels. Converted to kilograms and hectares, this means 2.03 hectares/ton in Germany and 4.83 hectares/ton in Russia. As wheat imports from these two countries dominate the statistics and are of similar proportions in 1850, I have entered their average = 3.43 hectares per ton.

7 Mitchell (1962:180) provides the following figures for imports of raw cotton from USA to the United Kingdom: 1815 = 54.000.000 lbs = 24,494 tons; 1850 = 493,000,000 lbs = 223,623 tons.

8 Mitchell (1962:190) estimates the average domestic wool clip of the United Kingdom in 1800–1819 at 100,000,000 lbs = 45,360 tons and in 1850–1854 at 135,000,000 lbs = 61,236 tons.

9 Clark (1993:228) estimates the output of wheat in Britain in 1851 at 12.6 bushels per acre. Converted to kilograms and hectares, this gives 1.14 hectares per ton.

10 Schumpeter (1960:59) for 1800 estimates 55,586,341 lbs = 25,213 tons. Mitchell (1962:180) gives 1815 = 101,000,000 lbs = 45,813 tons; 1850 = 664,000,000 lbs = 254,921 tons. Schlote (1952:21) gives 396,131 tons for 1854.

11 Mitchell (1962:191) provides the following figures for raw wool imports to Great Britain in 1800: 8,418,000 lbs = 3,818 tons. The remaining figures (ibid., 192) are for the United Kingdom, with re-exports subtracted: 1820 = 9,700,000 lbs = 4,399 tons; 1830 = 31,600,000 lbs = 14,333 tons; 1850 = 59,900,000 lbs = 27,170 tons. Schlote (1952:21) gives 47,376 tons in 1854.

12 Mitchell (1962:98) gives net imports of wheat and wheat meal and flour (with exports and re-exports deducted) to the United Kingdom in 1850 as 19,965,000 cwt = 1,014,222 tons. However, the sum of principal imports in the same source (ibid., 100) amounts to no more than 6,849,000 cwt = 347,929 tons, which leads me to suspect that the former figure in fact applies to thousands of quarters (19,965,000 quarters = 253,555 tons). Schlote (1952:61) gives the annual average wheat imports of the United Kingdom in 1849–1854 as 861,790 tons. In view of the contradictory nature of Mitchell's figures, I have used Schlote's.

13 Farnie (1979:199) estimates output of yarn and cloth, respectively, per hand for the following periods: 1819–1821 = 968 lbs and 322.5 lbs; 1829–1831 = 1,546 lbs and 520.7 lbs; 1849–1851 = 3,079 lbs and 2,437.8 lbs. I have summed up the total output and divided by the total number of hands employed in yarn and cloth production, yielding the following figures per employed: 1819–1821 = 1,290 lbs = 0.58 ton; 1829–1831 = 2,066 lbs = 0.93 ton; 1849–1851 = 5,516 lbs = 2.5 tons. I have estimated the number of annual working hours per employed at 3,000, based on six ten-hour days per week and fifty weeks per year. This would give the following number of hours per ton of cotton manufactures produced: 1819–1821 = 5,172; 1829–1831 = 3,225; 1849–1851 = 1,200.

14 Mitchell (1962:199) gives the number of employed in the wool industry in the United Kingdom in 1850 as 154,000. Assuming fifty sixty-hour weeks, the total number of annual working-hours can be approximated as 462,000,000. When this figure is divided by the number of tons of woollen manufactures produced (66,303), the hours of labour required for production of one ton of wollen manufactures can be calculated as 6,968.

15 Clark (1993:228) estimates the output of wheat per worker in Britain in 1851 as 272 bushels = 7,616 kg. Assuming 3,000 hours of work per year, this means 394 hours per ton.

16 According to Blaug (1961:377), quantity of cotton yarn produced can be derived by deducting loss of weight in spinning from consumption of raw cotton, reckoned in the years 1828–1861 at 11 per cent. 1 ton × 89 per cent = 890 kg.

17 Deane (1957:211) quotes an assessment from 1741 that roughly one-quarter of the weight of raw wool is lost in washing and processing.

continued

18 Schlote (1952:61) estimates the annual average wheat harvest in Great Britain in 1849–1854 at 3,620,000 tons.

19 I have used the figure for tons of raw cotton imported (254,921 tons) reduced by 11 per cent (cf. Blaug 1961:377) to estimate yarn production at 226,879 tons. I have then deducted the export of yarn, estimated at 24.9 per cent or 56,946 tons, to approximate the production of cloth at 170,050 tons.

20 According to Farnie (1979:7), the average annual volume of exports of cotton yarn and cloth in 1839–1841 was 254,100,000 lbs = 115,213 tons. The mean annual rate of increase in exports 1840–1872 is given as 3.9 per cent (ibid.), which in 1850 would mean around 168,910 tons. By deducting the proportion of yarn (56,946 tons) I have arrived at the figure 111,964 tons of cotton cloth exported 1850.

21 For 1741, Deane (1957:211) gives 262,175 packs of 240 lbs or 62,922,000 lbs = 28,541 tons. For 1799 and 1805, I have calculated 75 per cent of raw wool consumed in England and Wales, including imports. Deane (ibid., 220) gives the following figures for wool consumed (including imports): 1799 = 98,000,000 lbs = 44,452 tons × 75 per cent = 33,399 tons; 1805 = 102,000,000 lbs = 46,267 tons × 75 per cent = 34,700 tons. For 1850, I have added raw wool produced (61,235 tons) and imported (27,170 tons) and multiplied the total (88,405) by 75 per cent, yielding 66,303 tons.

22 Thompson (1968:75) gives the average annual imports of guano to Britain in 1851–1853 as 165,350 tons.

23 Mitchell (1962:291) gives the value of raw cotton imported to Great Britain in 1850 as £21,532,000. Divided by the number of tons imported the same year (254,921), this would mean a price of £84.46 per ton. Schlote (1952:21) gives the price of raw cotton in 1854 as £50.94 per ton.

24 Mitchell (1962:291) gives the value of imported wool in 1850 as £1,953,000. Divided by the number of tons imported in the same year (27,170), this would mean £71.88 per ton. Schlote (1952:21) gives the price of imported wool in 1854 as £137.18 per ton and (ibid., 116) the average price of wool in 1854–1860 as £7 per 100 lbs = £154 per ton.

25 Thompson (1968:76) gives the average annual value of imported guano in 1851–1853 as £2,628,000. Divided by the number of tons imported (165.350), this gives an exchange value of £15.89 per ton.

26 Drawing on Farnie's (1979:10) figures for average annual value of exports of cotton manufactures (£18,733,000 in 1814–1816) and mean annual rate of increase (2.9 per cent between 1819–1821 and 1859–1861), I have calculated the total value of cotton exports in 1840 as £37,202,000 and in 1850 as £49,513,000. By dividing these figures with the number of tons of cotton manufactures exported (115,213 tons in 1840 and 168,910 tons in 1850), I have estimated the exchange value of one ton of cotton manufactures exported at £322 in 1840 and £293 in 1850.

27 For 1805, Deane (1957:219) estimates a total of £18,500,000 for the final sales value of woollen manufactures in England and Wales. Divided by the number of tons produced (34,699), this gives an exchange value of £533 per ton. Deane (ibid., 218–219) suggests an overall multiplier of 2.5 to estimate the value added to raw wool in the woollen and worsted industries of the early nineteenth century. Applied to Schlote's (1952:21) estimate of the price of imported wool in 1854 (£137 per ton), it would give the figure £342 per ton of woollen manufactures. Using his figure for the price of domestic wool (£0.154/kg) in 1854–1860 (ibid., 116), however, we get £385 per ton.

28 Schlote (1952:61) estimates the average home price of wheat in the United Kingdom in 1849–1854 as 48.2 shillings per quarter. Converting 1 quarter = 12.7 kg, this would mean £222 per ton.

Table 5.2 Estimates of the purchasing power of £1,000 around 1850 in terms of quantities of some key commodities and their embodied inputs of labour and land

Commodity	*Volume for £1,000*	*Embodied labour*	*Embodied land*
Raw cotton	11.84 tons	32,619 h	58.6 ha
Cotton manufactures	3.41 tons	4,092 h	1 ha[1]
Raw wool	13.91 tons	13,743 h	385.3 ha
Woollen manufactures	2.92 tons	20,346 h	1 ha
Wheat (domestic)	4.5 tons[2]	1,773 h	5.1 ha
Wheat (imported)	4.5 tons	5,148 h	15.4 ha
Guano	62.93 tons		

Notes
1 The figure should probably be less than one hectare, i.e. the space occupied by a factory.
2 I am assuming that prices of domestic and imported wheat are similar.

These figures lead to several observations on the asymmetric exchanges of embodied labour and land that contributed to the accumulation of industrial capital in nineteenth-century Britain. Let us suggest a couple of examples:

a In exchanging, on the world market in 1850, £1,000 worth of cotton manufactures for £1,000 worth of raw cotton, Britain gained almost 800 per cent in terms of embodied labour and about sixty times the space utilized. These gains, representing technological superiority (i.e. capital) accumulated over several decades, further increased as profits were invested in continued technological development. (The efficiency of labour thirty years earlier had been 5,172 hours per produced ton of cotton manufactures, which means a 331 per cent increase in real output per hour between 1820 and 1850.[4])

b In exhanging £1,000 worth of woollen manufactures for £1,000 worth of imported raw wool, Britain in 1850 lost 32 per cent in terms of embodied labour but gained almost 400 times the space utilized.

3 We can also use these figures to illuminate the role of industrial export production as a means of environmental load displacement:

a To export £1,000 worth of cotton manufactures from Britain implied much less of a drain on domestic resources than to export £1,000 worth of woollens produced from domestic wool, namely only 20 per cent of the embodied labour and virtually none of the embodied land. To import the raw cotton to generate £1,000 of income from textile production rather than rely on domestic wool meant saving 16,254 British work-hours[5] at the expense of 11,626 work-hours in America and saving 107 British hectares at the expense of twenty hectares in America.[6]

b To import £1,000 worth of wheat from Russia and Prussia rather than growing it at home meant saving 1,773 British work-hours at the expense of 5,148 foreign work-hours and 5.1 British hectares at the expense of 15.4 hectares in Russia and Prussia. This does not take into account, however, the time–space demands of the domestic export industry whose profits paid for wheat imports. In terms of the ratio of embodied land and labour to exchange value, neither cotton nor woollen manufactures could compete with wheat, but here we see how the exigencies of demography inevitably constrain economic rationality. The growing population of Britain had first and foremost to be fed, and British wheat production around 1850 (3,620,000 tons × 1.14 hectares) demanded 4,126,800 hectares of the best agricultural land in the country. To spend £94 (1.12 × £84.46) on raw material and 4,092 work-hours in the cotton industry in order to be able to import 4.5 tons of wheat (the equivalent of only 1,773 work-hours but crucially also of 5.1 hectares prime farmland in Britain) – or 63 tons of guano – is economically rational under conditions of land scarcity, as was very obviously the case in early nineteenth-century Britain. In exchanging cotton manufactures for foreign wheat, rather than grow the wheat at home, Britain lost over 56 per cent in hours of labour but gained almost the entire land area used to cultivate it.

c Most centrally, from the perspective sketched at the outset in this chapter, the displacement of demands on land represented by the appropriation of the productivity of 1.1 million hectares of cotton fields in North America, which in terms of revenue ultimately generated £66,475,547, meant the liberation of the over six million hectares[7] in Britain that would have been required to generate the equivalent amount of revenue from woollen manufactures. This land area exceeds by almost 50 per cent the total area of British wheat cultivation (4,126,800 hectares) in 1850. If we add to this the imports of wool from overseas, an additional 0.75 million British hectares were set free, e.g. to fuel the labour force.

d This kind of analysis certainly deserves to be expanded to cover the other major commodities that provided the global framework for British industrialization. This applies, for instance, to those periods when Britain relied on imports of iron from Scandinavia, the production of which required huge volumes of Scandinavian labour and charcoal. Other important commodities embodying significant amounts of labour and land in the periphery in the mid-nineteenth century include timber, sugar, tea, coffee, wine, silk, flax, vegetable oils, dyes, hemp, and tobacco.

e A further elaboration (or simplification) of this argument would be to demonstrate that even the flows of embodied labour can ultimately be converted into flows of embodied land. If the embodied hours of labour are converted into man-years and multiplied by an average ecological footprint for this category of labourers, commodity flows could conceivably be analysed exclusively in terms of the exchange of space. A brief calculation indicates that such a translation of (labour) time into space would, in the case of mid-nineteenth-century England, further accentuate the unequal global exchange of embodied land, even if the per capita footprints of British factory workers in 1850 are assumed to be twice as large as those of plantation workers in the colonies. If it required about 12.8 North American man-years to produce £1,000 worth of raw cotton, but about 1.6 British man-years to produce £1,000 worth of cotton textiles, and the average ecological footprints of these two categories of labourers are (very hypothetically) estimated at 0.5 and 1.0 hectares, respectively, the exchange of labour embodied in these volumes of commodities would still represent an exchange of 1.6 hectares of embodied British land for 6.4 hectares of embodied American land. This conclusion presupposes that most of the land requirements of British and American labourers at this time were provided domestically (i.e. from within the respective nation).

It is obvious that only a few of Britain's imports in the mid-nineteenth century actually in themselves contributed exergy or negative entropy to its emerging industrial infrastructure. In terms of thermodynamics, as Britain was then self-sufficient in (and a net exporter of) fossil fuels, the prime candidates were wheat and sugar, which provided food energy for a significant proportion of the labour force (Pomeranz 2000:313–314). In terms of materials, the prime candidates were iron and timber. The thermodynamic perspective that I have advocated, however, has little regard for national boundaries, as long as imports of exergy and exports of entropy have not developed into national interests, as in the current struggles over oil (Klare 2001) or deliberations about emission permits (Martinez-Alier 2002). Unequal exchange can be a very local affair. As the world's first industrial districts were emerging in eighteenth-century Yorkshire and Lancashire, the most significant extractive periphery may have been no further off than the nearest coalfields, where the severity of working conditions and environmental degradation may well have equalled that of the Alabama cotton plantations (Hobsbawm 1968:252–256, 281, diagram 5b). During the course of the nineteenth century, this periphery expanded to truly global proportions, but the logic remains the same to this day. Although deeply submerged beneath statistics catering to the concerns of mainstream economics, this logic remains founded on the imperative – for all kinds of complex systems or structures – to maintain a net input of negative entropy. This can be achieved through a variety of social strategies complementing and succeeding each other over the course of history. From this perspective, we have seen how the import of raw materials for the textile industry, although not in itself representing an energy source, served to make room for the energy provisioning of the British labour force. Factory workers and coalminers alike ran on wheat.

The British textile industry for two centuries played a central role in generating revenue to maintain an expanding technological infrastructure, originally as a means of coping with severe shortage of land (Wilkinson 1973). As this expansion continued and spread to continental neighbours and former colonies, mainstream discourse has conspired to ignore the unequal exchange of (labour) time and (natural) space on which technological development has been founded from the start. The complex webs of the modern world economy no doubt make analyses such as these extremely difficult, but we can remain quite certain that capital, however much it tries, will never be able to delink itself from labour and land. The rationale of machine technology is to (locally) save or liberate time and space, but crucially at the expense of time and space consumed elsewhere in the social system. The general hypothesis offered here is that to save time and space by the application of increasingly efficient technologies may often tend to imply that someone else in the world system is *losing* time or space in the process. The process of globalization that was in full swing in the early days of industrial capitalism then as now relied on what Harvey (e.g. 1996) has called 'time–space compression'. The point that I have wanted to make here is that such analyses of the very essence of technological development need to be founded on a concrete, empirical understanding of what I would call time–space *appropriation*.

Rethinking the Industrial Revolution: implications for writing world history

In the remainder of this chapter I want to explore the role of the concept of technology in the way we write world history. Over the past two decades, several persuasive critiques of Eurocentric narratives, which explain the 'rise of the West' in terms of uniquely European conditions, have offered perspectives emphasizing the global conjunctures and contingencies that provided opportunities for European expansion (see particularly Blaut 1993, 2000; Frank 1998; Pomeranz 2000). The latter way of writing world history is one with which most anthropologists like myself would concur, because it relativizes cultural historiographies and deconstructs the essentialist legacy of Eurocentrism, which can be traced back to colonial ideologies and even racism. In fact, my agreement with this shift of perspective extends to the point where I suspect that it has not been radical enough. In scrutinizing the arguments challenging Eurocentrism, I have found that the aspect of Western cosmology that has been most resistant to deconstruction is the concept of *technology*. I thus want to focus this brief discussion on the way technology is treated in the seminal works of James M. Blaut, Andre Gunder Frank, and Kenneth Pomeranz.

Blaut (1993, 2000) and Frank (1998) derisively review the modern history of Eurocentric historiography from Adam Smith, Karl Marx, and Max Weber to Lynn White, Robert Brenner, and Eric Jones. The shift of perspective that they propose, building on earlier critics such as Eric Williams, Janet Abu-Lughod, and Edward Said, is incontrovertible. The economic and technological expansion of Europe in the nineteenth century cannot be explained in terms of conditions or characteristics specific to Europe, whether cultural or environmental. Such accounts, referring to allegedly unique European features such as 'rationality' or 'inventiveness', commit the error of essentializing what appear to be geographically circumscribed developments into local (i.e. national or continental) accomplishments, rather than recognizing their emergence as conditioned by global processes. The latter recognition is fundamental to world-system awareness and uncontroversial enough not to require recapitulation. Instead, we shall look more closely at the specific role of technology in these non-Eurocentric accounts. Since a pivotal phenomenon debated in these deliberations is the so-called Industrial Revolution itself, it should indeed be significant to scrutinize mainstream understandings of technology.

Even if non-Eurocentric historiographers have managed to extricate themselves from discursive assumptions about European superiority, I shall argue that they may not have managed to abandon the specific definition of 'technological progress' that is equally diagnostic of the modernist world-view (cf. Adas 1989, 2006; Marsden and Smith 2005; Friedel 2007; Headrick 2010). The very materiality of technological superiority apparently makes it very difficult to deconstruct. Contemplating technology, even as profound a cultural theorist as Marshall Sahlins (1976:207) concedes that 'thought can only kneel before the absolute sovereignty of the physical world'. J. M. Blaut (2000:38) observes:

> It is not hard at all to prove that one tool produces more food, or more cloth, or more enemy casualties than another. Hence the peculiarly convincing quality of arguments grounded in technology; and hence the preference for this kind of argument, and evidence, in efforts to prove the superiority of Europe at all times...

As studies by Bruno Latour (1993) and others suggest, however, technological objects tend to embody social relations from which they cannot be extricated in any simple manner. The understanding that artefacts such as commodities are in some sense *social* goes back at least to Karl Marx, who observed that relations between people in capitalist society assume the appearance of relations between things. It is thus paradoxical that the notion of 'technological progress', suggesting that technological objects can be arranged along a normative scale of generalized improvement, continues to constrain mainstream discourse on development and history. We know that the most 'advanced' technologies may have disastrous social, political, or ecological implications at a global scale – not least in terms of the shifting of burdens to other populations – yet we rarely allow such insights to contaminate our assumptions about technological progress. Rather than view modern, capitalist technology as a specific social strategy to accumulate power, through the appropriation of (human) time and (natural) space from other people – i.e. as a strategy of elite *capacitation* – the technology itself tends to be viewed as exempt from political critique (Hornborg 1992, 2001b). This is particularly evident in Marxist discourse, which acknowledges that machinery represents the exploitative accumulation of appropriated labour time but simultaneously glorifies technology as the road to global proletarian emancipation. A similar contradiction pervades the discourse on ecological sustainability, which generally assumes that the environmental degradation generated by technology will be alleviated by more 'technological progress'. I hope to illustrate how even non-Eurocentric world-system analysts, while earnestly struggling to dethrone Europe and the 'West' as intrinsically generative of 'technological progress', tend to face difficulties reframing such progress in terms of global power relations. Even if an historian such as Blaut (2000:42) is able to acknowledge how the 'modern belief in "perpetual progress" ' is a product of 'the growth of capitalism' – i.e. a reconceptualization of time within a privileged social space – the conceptual hurdle still remains of dismantling the Cartesian immunity of technology to political critique.[8]

The general argument about technology pursued by Blaut (1993, 2000), Frank (1998), and Pomeranz (2000) is that Europeans by the late eighteenth century were not more inventive than people in other core areas of the Old World, such as China or India, and that the technological inventions that have previously been attributed to Europeans were generally invented elsewhere. This is of course a valid way of countering the converse claims of Eurocentric historians of technology such as Lynn White, Eric Jones, and David Landes, but deliberations about the priority of inventions do not address the more fundamental question of what technology *is*. Only briefly do Blaut and Frank touch upon this latter issue, but their intuitions about the world-systemic nature of modern technology deserve to be elaborated. Rather than view technological capacity as an essentialized attribute of a given population, such an elaboration would entail understanding the specific social application of a technology as a kind of 'field phenomenon' emerging at certain points in the global system, as an expression of processes involving the system as a whole.

Each of the non-Eurocentric historians I have mentioned devote specific sections of their texts to the topic of technology in world history, where their approach to the matter is distilled (Blaut 1993:108–119, 2000:31–42; Frank 1998:185–205; Pomeranz 2000:43–68). Blaut is obviously frustrated by the power of technological explanations of European dominance: 'Among all of the narrow-minded ways of looking at history, technological determinism is the one most congenial to Eurocentric tunnel vision. It has the appearance, the illusion, of cold-blooded scientific fact' (1993:108, 2000:38). He then turns immediately to the question of priority, emphasizing that Europe did not 'advance technologically' beyond other continents until the Industrial Revolution in the eighteenth century. The main point here seems to be to show that technological inventiveness was not a specifically European feature that could be traced to the Middle Ages. In an argument largely directed at the work of Lynn White (1962), Blaut (1993:110–117, 2000:31–38) lists allegedly European, pre-industrial inventions that were in fact derived from, or at least equally developed, elsewhere: iron-working (Middle East, West Africa); ploughs (Middle East, China); the horse collar (Eurasia); textile production (China); mechanical clocks (China); shipbuilding (China); cannons (China); printing with movable metal types (Korea); etc. Blaut (1993:115) concludes: 'It was only after 1492, with its utterly revolutionary consequences, that European technology acquired the beginning of an edge over Asian and African.' In other words, this technological 'edge' only began to crystallize as a consequence of the consolidation of the modern world system.

This is precisely also the conclusion of Andre Gunder Frank (1998). Citing Robert Adams (1996) and others, Frank observes that technological developments in Europe had very little to do with a European propensity for science: indeed, technology contributed more to the advancement of science than the reverse (Frank 1998:191). Like Blaut, he goes on to list a number of technologies in which, if anything, early modern Europe lagged behind other core areas of the Old World: shipbuilding, printing, textile production, metallurgy, overland transports, etc. Frank (ibid., 197) takes an important step toward de-essentializing technological capacity when he notes that 'armaments technology was rapidly diffused to *anyone in a position to pay for it*' (italics added). For Frank, the expansion of Europe was founded on the geographical and historical accidents that gave it access to the New World, and particularly its silver, which (temporarily) shifted the balance of global economic power from the East to the West. For our present purposes, his most important conclusion is that 'there was no *European* technology!' (ibid., 204). Technological inventions are not attributes of specific populations, but capacities that are rapidly adopted by any population that can afford them. 'That is', he continues, 'technological development was a *world economic process*, which took place in and because of the structure of the world economy/system itself.' This is the closest a world historian has yet come to radically reconceptualizing technology as a total socio-ecological phenomenon. To fully appreciate the far-reaching implications of this perspective, we can consider Frank's (ibid., 333–334) own summary of his understanding of the Industrial Revolution and the 'rise of the West':

> We have seen that most efforts to address this question have heretofore suffered from misplaced concreteness, because they looked for these factors only under the European street light. Yet since the West was part and parcel of the global world economy, the West could not rise on its own or by itself. Instead, any such Western rise must have been *within* the world economy itself. Therefore, it is useless to look for the causes of this rise only or even primarily in the West or in any part of it – unless the 'use' for doing so is only ideological, that is to pat oneself on the back and put all others down as incompetent.

Like Blaut and Frank, Pomeranz (2000:43–48) shows that Europe in 1750 was not 'technologically superior' to, for instance, China, India, or Japan, but was still struggling to 'catch up' with these areas in fields such as agriculture, textile production, porcelain, medicine, sanitation, and even iron-making.[9] The relevant question is why Europe by the nineteenth century had absorbed what Pomeranz refers to as the 'best available technologies' developed elsewhere in the world. There was obviously something about this particular geographical region at this point in time that encouraged the adoption and refinement of mechanical devices capable of increasing production per invested unit of domestic labour and land. If this was the essence of the Industrial Revolution, a world-systemic perspective should prompt us to ask not only whence these inventions derived, but more importantly what were the global consequences of the European strategies for saving labour and land to which they were applied. Considering their implications for populations and ecosystems in Africa, Asia, and the Americas, we may need to nuance our definition of 'technological progress'. Were Europe's industrial technologies in the nineteenth century 'best' not only for Europe, but for the world as a whole? The answer will inevitably appear different in a so-called 'full world', where it is arguable that technological *rationality* can no longer be geared to the economic success of nations ruthlessly competing for market shares, but should reflect a concern with globalized humankind managing finite resources and living sustainably on a single planet.

Although it has generally been argued that the rationale of the Industrial Revolution was the application of labour-saving innovations that enabled Europe to oust competitors like the Indian textile industry, Pomeranz (2000:49–68) argues that the most important innovations were in fact *land*-saving ones. Much as Richard Wilkinson suggested in *Poverty and Progress* (1973), the so-called 'progress' or 'development' of the Industrial Revolution was prompted by the cul-de-sac of ecological constraints that England shared with other populous core areas of the Old World in the eighteenth century, but that it was uniquely able to transcend through its access to the Americas and its own deposits of coal. In displacing their environmental demands to other continents and other ages,[10] industrialized societies have from the start pursued a radically different strategy for dealing with the 'land constraint' than, for example, eighteenth-century China. Although Pomeranz (2000:264) characterizes this strategy as 'abolishing' the land constraint, the 'ecological relief' that Europe gained through the Industrial Revolution occurred at the expense of other populations, ecosystems, and generations.[11]

Pomeranz (2000:52) concedes that 'the application of coal and steam power to all sorts of processes eventually led to enormous labour savings', but proposes that '(i)f the makers of the Industrial Revolution were primarily economizing on expensive labour, they were unaware of it'. They were rather struggling with the 'bottleneck' represented by constraints on the expansion of textile production. Pomeranz (ibid., 55) observes that 'it is unlikely that the necessary land to relieve this bottleneck could have been found in Europe'. The fibre needed for textiles needed land, and 'competition for land among Malthus's four necessities – food, fuel, fiber, and building materials – was growing ever more intense in much of eighteenth-century Europe' (ibid., 56). Soil degradation, deforestation, and overgrazing in western Europe at this time brought erosion, dust storms, declining yields, waterlogging, acidification, and unfavourable climate change. Even if such ecological constraints were common to other core areas in the Old World, which all imported land-intensive commodities in exchange for manufactures (particularly textiles), Europe was able to use the New World and the Atlantic slave trade to turn 'manufactured goods created without much use of British land…into ever-increasing amounts of land-intensive food and fiber (and later timber) at reasonable (and even falling) prices' (ibid., 242–243, 269). Not only did the Americas offer Europe vast areas of conveniently depopulated land, but also huge quantities of silver with which to purchase additional 'ecological relief' from the rest of the Old World (ibid., 274).

Pomeranz (2000:274–278, 313–315) provides tangible measurements of how, in various ways, the Industrial Revolution brought 'ecological relief' to nineteenth-century Britain. Citing Mintz (1985), he observes that to replace the calories from sugar consumed in the United Kingdom in 1831 with food energy from domestic harvests would have required up to 2.6 million acres of average-yielding English farmland. Raising the sheep to replace the cotton fibres imported in 1830 would have required over twenty-three million acres of pasture and hay, a figure which by itself 'surpasses Britain's total crop and pasture land combined'.[12] Pomeranz calculates that imports of Baltic and American timber in the early nineteenth century substituted for the output of over 1.6 million acres of British woodland, and cites E. A. Wrigley's (1988) calculation that England's annual use of energy from coal already in 1815 would have been equivalent to at least fifteen million acres of woodland (Pomeranz 2000: 59).[13] These land-saving strategies grew tremendously in significance through the nineteenth century. Britain's sugar imports increased eleven-fold from 1815 to 1900, its coal output fourteen-fold, and its cotton imports 'a stunning twenty-fold' (ibid., 283–284). What these three commodities alone meant in terms of millions of acres of 'ecological relief' is not difficult to estimate,[14] but Britain over the same period also imported huge quantities of American grain, beef, lumber, and other land-intensive products.

Pomeranz (2000:282–283) concludes that

> it seems likely that the exploitation of the New World, and of the Africans taken there to work, mattered in many ways above and beyond those reflected in our ghost acreage figures. Taking all the indices together, it seems likely that this exploitation did more to differentiate western Europe from other Old World cores than any of the supposed advantages over these other regions generated by … institutions within Europe.

In other words, 'forces outside the market and conjunctures beyond Europe deserve a central place in explaining why western Europe's otherwise largely unexceptional core achieved unique breakthroughs and wound up as the privileged center of the nineteenth century's new world economy' (ibid., 297).

This reinterpretation of world history since the sixteenth century, proposed by Blaut, Frank, and Pomeranz, has implications far beyond the abandonment of Euroexceptionalism. To reconceptualize the material artefacts that surround us as crystallized relations of unequal exchange would be to pursue a fundamentally Marxian understanding of capital accumulation beyond the Cartesian boundaries that have hitherto kept technology itself exempt from political critique.

6 Value, unequal exchange, and uneven development

The main argument in this book is that the economic, technological, and environmental polarizations of global society can be understood in terms of asymmetric transfers of resources that are made invisible by the dominant ways of representing development, economic growth, and technological progress. The acknowledgement of such asymmetric transfers – what I have called the 'thermodynamics of imperialism' or 'ecologically unequal exchange' (Hornborg 1992, 1998b) – is fundamental to understanding not only development gaps, but the very phenomenon of 'technology' as a social redistribution of resources. In recent years, it has been gratifying to see an increasing number of researchers involved in defining and measuring different kinds of ecologically unequal exchange (cf. Jorgenson and Clark 2009), even if the implications of this work for a radical reconceptualization of technology remain difficult for most of us to digest.

In this chapter, I will take a closer look at the concept of unequal exchange, which is a cornerstone not only of Marxian social theory but also of much ecological and post-colonial critique of the notion of development. Many social scientists, looking at the world around them, are intuitively convinced that there is such a thing as 'unequal exchange' but would admit to having a hard time defining it. The problem of unequal exchange is a paradigmatically Marxian topic in that *our difficulties in conceptualizing it can be seen as part of the conditions for its existence.* Thus it cannot be understood other than through an analytically demanding combination of epistemological and ontological arguments that require at different steps in the analysis the approaches of both deconstruction and objectivism.

My previous attempts to conceptualize ecologically unequal exchange (e.g. Hornborg 1992, 1998b, 2001b) have raised two related kinds of objections that deserve response. The first is that the very notion of 'unequal' exchange must imply some kind of value judgement, the second that it should refer to the specific definition applied by Arghiri Emmanuel

(1972). In the following section, I hope to show why the first objection must be rejected, by arguing that if objectively quantifiable net transfers of resources can be shown to be conducive to uneven capital accumulation (or development), we do not need a normative concept of value or inequality in order to observe that uneven development is a result of unequal exchange. (This important argument elaborates some observations I made toward the end of Chapter 4.) In the second part of this chapter, I shall address the second objection.

How to define unequal exchange without recourse to a normative theory of value

Few mainstream economists today would recognize the notion of unequal exchange as an acceptable category of economics, but tend to deal with the problem of global inequalities by referring to monopolies and imperfect information. The economists' solution is to try to envisage conditions for completely free trade and more perfect competition and information flows, but if after two centuries the supposedly equalizing doctrines of free market economics continue to remain a distant mirage, it should be incumbent on economists to devise more realistic strategies to achieve equality. Suffice it to say here that as long as exchange is conducted in terms of monetary exchange values, and prices are understood to reflect the rational or even benevolent logic of market forces, there is no way – other than under conditions of monopoly – that a market transaction can be classified as unequal. A million dollars' worth of Swedish Volvos exchanged for a million dollars' worth of Venezuelan oil is *by definition* perfectly equal in terms of exchange value, which is the only gauge that neoclassical economic theory is capable of applying. However profoundly we manage to deconstruct the phenomenon of money as a vacuous, semiotic delusion, aptly classified by Marx as a species of fetishism, the ideological and practical hegemony of exchange value, gauged in terms of market prices, remains more intact than ever. The foundations of modern economics were devised by and for British bankers and stock traders in the early nineteenth century, yet continue globally to pervade the lives and thoughts of dominator and dominated alike.

Initially influenced by the Physiocrats' conviction that land was the ultimate source of value (Gudeman 1986), David Ricardo later subscribed to a labour theory of value that also became fundamental to the ideas of Karl Marx. Marxian theory has from the very start struggled with the analytical problem of how to effectively challenge the mainstream trust in money and in the fairness of market logic. Marx suggested that the market price of labour did not do justice to its 'real' value. Although he and his followers would be the last to admit it, he thus offered what I would call a *normative* theory of value in the sense that it defined value not in terms of the actual, subjective valuations of market actors – as in the fundamentally descriptive, neoclassic notion of 'utility', reflected in prices – but in terms of an analytical construct (the labour theory of value) that itself proposed to define an *objective basis of value*. Rather than accepting value as contingent on the aggregated and transient subjectivities of consumers, Marx's highly justified struggle to uncover the material conditions of accumulation (and the obvious exploitation of the working class) thus led him to conceive an analytical oxymoron. Beyond the mystifying price tags on labour that we know as wages, he pursued an objective foundation of value. Following Ricardo, Marx believed that embodied labour value was systematically reflected in exchange value. But valuation is a subjective act, and to 'objectively' define value is (paradoxically) itself an act of valuation.

Although meant to serve a commendable political purpose in Marx's own time, this approach to value must be rejected as analytically untenable. Marx realized that in order to challenge the market ideology legitimizing capital accumulation, e.g. by positing asymmetric transfers of surplus value, it would be necessary to acknowledge some other gauge than price, but unfortunately chose to conceive this other gauge in terms of value. The mistake was to conceptualize material asymmetries in terms of a subjectivist terminology. The concept of value is itself normative. I doubt that it will ever be possible to convince economists or market actors that academics have a better knowledge of the 'real' value of things than the majority of market actors themselves. So, what does this other gauge – labour – represent, if not value?

The answer may be easier to detect if we turn to another kind of normative theories of value, namely those that underneath the price tags recognize not primarily labour but more generally *energy*. There have been many proponents of such theories over the years, including the so-called Technocrats in the 1930s and, more recently, the ecologist Howard T. Odum (1988), the economist Robert Costanza (1980), and the sociologist Stephen Bunker (1985). Odum's notion of *emergy* (or 'energy memory') echoes Marx in suggesting that the 'real' value of a commodity reflects the amount of energy that has been invested in its production. Like Marx, Odum used an ultimately normative theory of value to pursue a putatively scientific argument that exchange can be viewed as unequal in the sense that some social categories are not properly compensated (Odum and Arding 1991). Costanza (1980), on the other hand, traced empirical correlations between embodied energy and price, in effect offering a *descriptive* energy theory of value, without considering the possibility of unequal exchange resulting from discrepancies between prices/wages and energy 'value'.

What both labour and energy have in common is that they are measures of *productive potential*. They are literally the 'productive forces' of any production process. But, contrary to Marx and Odum, there is no specifiable relation between the amount of productive potential that has been invested in a commodity and the way it will be evaluated on the market. Rather than reduce economics to thermodynamics, our task should be to see how the two are related. We need to keep them analytically separate while showing how they are interfused in actual social processes.

We have no theoretical reason to posit a connection between the attractiveness of commodities and the volume of labour time (or any other biophysical metric) that has been invested in their production. The former is tantamount to value (utility = exchange value = price), the latter one of several possible measures of embodied productive potential (also including, for example, energy, eco-productive land area, volume of materials, etc.), and there need not be any positive correlation between them. The so-called 'transformation problem' is thus a non-problem. Value is what consumers desire. To claim that value is essentially a question of invested labour time (or energy, or land area, and so on) is itself a valuation, in other words, to confuse that which is to be explained with the theory purporting to explain it.

Rather than say that we as social or natural scientists have access to a more authentic measure of value than the people who do the valuing, we here have reasons to agree with the mainstream economists that value is defined by the cultural preferences of consumers. This agrees well with anthropological and sociological studies of the semiotics of consumption following the work of Jean Baudrillard (1972), Marshall Sahlins (1976), and Pierre Bourdieu (1984), who all argued that valuation occurs within the specific cultural logic subscribed to by some particular group of people. Theories of value should be descriptive, i.e. they should be based on the valuations that people actually make, not on what theorists claim to be an objective source of value. Normative theories of value make the mistake of inserting themselves on the same logical level as the phenomena they are to explain.

How, then, can we posit the occurrence of unequal exchange without recourse to a normative theory of value? This can be done by analytically demonstrating that there is, in very general terms, a systematic relation between (1) flows of productive potential, (2) flows of utility or exchange value (price), and (3) economic growth and the accumulation of capital. But this relationship is not usefully expressed as Marx or Odum would have it, that investment of labour or energy somehow translates into exchange value. Rather, there is a kind of *inverse* relation between productive potential and price that follows with logical necessity from the juxtaposition of the Second Law of Thermodynamics and the social institution of market exchange. We know that energy is not so much invested as it is *dissipated* in a production process (Georgescu-Roegen 1971). Finished products must represent an increase in entropy compared to the resources from which they were produced, yet they must be priced higher. If we consider, longitudinally, the transformation of a given set of natural resources into an industrial product, Odum's measure of energy memory must necessarily correlate positively with utility or price, but objectively speaking, the amount of remaining available energy will be *negatively* correlated with price. As utility or price increases, there will be less of the original, available energy left. This means that industrial centres exporting high-utility commodities will automatically gain access to ever greater amounts of available energy from their hinterlands. The more energy they have dissipated today, the more new energy they will be able to buy – and dissipate – tomorrow. Unfortunately, the logic of this argument has often escaped my critics (e.g. Brolin 2006:262).

As I was developing these ideas on the 'thermodynamics of imperialism' in the late 1980s (Hornborg 1989, 1992), I had not yet encountered Bunker's (1985) important contribution on ecologically unequal exchange in Amazonia, which proved to contain several ideas with which I found myself in agreement, and some which seemed less useful (Hornborg 1998b). Although most of the transfer of available energy to industrial sectors is dissipated in production, and a small share returned to their hinterlands in the form of industrial products and waste, a significant part of it is invested in an expanding, industrial infrastructure, which through a self-reinforcing logic involving economies of scale (Bunker 1985) will tend to augment this process of accumulation and the unequal exchange of energy on which it is founded. This, of course, is a very different way of describing what the economists know as growth. An intensification of industrial production will generally mean more competitive prices, expanding market shares, and rising profits for industrial sectors, which in turn means more purchasing power with which to appropriate even greater amounts of energy and other resources from peripheral sectors. An intensification of natural resource extraction, on the contrary, will ultimately lead to local resource exhaustion and ecological degradation, prompting investments to be shifted elsewhere and truncating cumulative economic expansion (ibid.).

Note again that although referring to Odum and Bunker, this account of unequal exchange is not tantamount to an energy theory of value, but rather the opposite. Like Marx, Odum, and Bunker argued, it is necessary to refer to another gauge than prices for assessing the effects on development of market exchange, but unlike their work this account avoids the pitfall of trying to objectively define value. In *not* offering an alternative theory of value, we not only avoid having to systematically contradict the valuations that people actually make, but are also free to suggest additional gauges of productive potential that could be used alongside energy and labour to illuminate processes of unequal exchange. A well-documented such metric, for instance, is the study of material flows and biophysical trade balances (Fischer-Kowalski 1998; Pérez Rincón 2006; Weisz *et al.* 2006; Krausmann *et al.* 2009; Gierlinger 2010).

In the previous chapter I suggested that an additional such perspective, which integrates the concerns of Marxian and ecological economists, might be expressed as the unequal exchange of time and space. I am thinking here of time and space as human as well as productive resources. Human time can be saved as well as invested (as labour) in production, and the same goes for space (or land). When considered in relation to the fundamental rationale of most modern technology, this means that time and space can be redistributed in global society through unequal exchange. Most technology can be visualized as devices for 'saving' time or space: time by increasing velocity (e.g. railways, cars, airplanes) and space by intensifying the use of land (e.g. through high-rise buildings or modern agricultural machinery). What we seldom take into account is that this local saving of time and space is made possible precisely by the expenditure or loss of time and space elsewhere in the global system. Just to give an early and fairly simple example, railways in the 1840s may have saved time – and accessed more space – for those who could afford to use them, but obviously at the expense of the underpaid labour time of vast armies of miners, loggers, steelworkers, and railway workers, as well as of the equally underpaid natural spaces where clear-cuts and strip mines were all that remained of the landscapes that had to be sacrificed in the name of technological progress.

The unequal exchange of time has to a large extent already been exposed by Marxian theory, notably in the work of Arghiri Emmanuel (1972), who showed that low-wage countries have to export more products in exchange for a given volume of imports from high-wage countries than they would have needed to if the wage level had been uniform. Yet I doubt that Marxists themselves have fully grasped the implications of this analysis for our understanding of the very nature of modern technology. If machines from the very beginning of the Industrial Revolution can be visualized as devices for saving time for *some* at the expense of the time available to *others*, it would not make sense to view the development of the productive forces as a cornucopia promising to emancipate the global proletariat.

If we add the more recent recognition, for example in the notion of ecological footprints (Wackernagel and Rees 1996), that there is also an unequal exchange of space, such hopes of technological emancipation seem even more untenable. The Industrial Revolution was not so much an absolute emancipation from land constraints as the local accumulation of a capacity to export and redistribute such constraints in global society (cf. Wilkinson 1973; Pomeranz 2000). It did not dissolve (European) land constraints once and for all as much as it provided Europe with ways of appropriating the land resources of other continents. What the post-development world might teach us is that technological progress or growth may not be the cornucopia that Ricardo and Marx generally believed, but local expressions of a kind of global zero-sum game. And what this means in terms of our understanding of concrete technology as a thoroughly social construct is even harder to digest, because it means that a tangible piece of machinery like a tractor or railway engine would simply not be feasible were it not for the uneven ways in which human time and natural space are *priced* in global society (Hornborg 2001b). The high-tech sectors of global society presently celebrating their efficient use of time and space appear largely oblivious of the extent to which this efficiency has been made possible by exploiting vast investments of human time and natural space historically and presently made elsewhere in the world system. Although such connections are generally concealed from their sight by virtue of geography or the passing of time, developed sectors owe as much to slavery and ecological devastation as to genius and entrepreneurship.

Finally, as also suggested in the previous chapter, even the net transfers of embodied labour can be mathematically converted into embodied land, namely by multiplying embodied man-years by the average ecological footprints for the relevant category of labourers. This illustrates how, in biophysical reality, the relations between Ricardo's three factors of production (land, labour, and capital) are quite different from how they are conceived in mainstream economic models. Rather than being mutually substitutable, the three factors are asymmetrically related, with land the ultimate source of both labour and capital. Capital, from the start, was generated in the appropriation of other people's land and other people's labour.

I should emphasize again that I have been using the notion of unequal exchange not in the moral sense of not getting one's money's worth, but in the naturalistic or realist sense of an objectively asymmetric transfer of some biophysical quantity or metric, not usefully referred to as value, by which the productive capacity of one social group is augmented at the expense of that of another. My argument is that industrial capitalism is founded and dependent on such objective, net transfers of productive potential. It is thus not a moral argument at the level of analysis, but can of course issue in a moral argument when articulated with the observation that an asymmetric transfer (net import) of energy or embodied land to one region or social group is the basis of a self-reinforcing accumulation of technological superiority and power vis-à-vis other regions or social groups.

How to disentangle the concept of unequal exchange from earlier scholasticism

The second objection to my earlier work is that it would be useful to restrict the concept of unequal exchange to the specific way in which it was applied by Arghiri Emmanuel (1972), namely as a result of international differences in the price of labour. This view is fundamental to two of the PhD theses recently produced at my own department (Brolin 2006; Nordlund 2010). I would thus like to take this opportunity to address some conceptual issues that should be central to development studies, but my focus here on Brolin's and Nordlund's approaches to these issues serves only to illustrate the kinds of miscommunication that such matters frequently generate. Together, these two contributions raise a number of questions that will be recognized as pivotal concerns of development studies in general.

Whereas most contributors to the discussion tend to assume an inevitable connection between theories of value and theories of unequal exchange, I maintain that the two concepts must be kept analytically distinct (Hornborg 1998b, 2001b). Briefly, as I have argued in the first section, unequal exchange is not a normative category, whereas an objectivist notion of value (i.e. one not simply equated with price) *is*. Much of the confusion regarding value is necessarily highlighted in the ambition to integrate a Marxian concern with the unequal exchange of labour, as in Emmanuel's (1972) calculations, with the ecological economists' concern with the unequal exchange of energy or embodied land. Like most Marxists – and although he momentarily pauses to acknowledge global Malthusian constraints – Emmanuel was a strong proponent of economic and technological growth. To bring Marxian and ecological economics together in a single theoretical framework thus necessarily requires transcending some major differences in fundamental assumptions.

As Brolin (2006) shows in his detailed history of economic ideas, attempts to illuminate the operation of international exchange by defining sources of value can be traced through a series of paradigms ranging from mercantilists and Physiocrats to classical and neoclassical economists, Marxists, and ecological economists. With all due respect to the immense inputs of human time and intellectual energy invested in these deliberations over the centuries, I must conclude, from reading Brolin's thesis, that the scholastic obsession with a reified notion of value in the eighteenth, nineteenth, and twentieth centuries is reminiscent, both in terms of scholarly output and ideological significance, of medieval theology. As any exegesis, such reviews will demand spending inordinate efforts on unravelling the contradictions and inconsistencies of individual scholars. While the history of such scholastic debates can be revealing, they contribute little to demystifying the glaring inequities and ecological devastation of the modern world.

Emmanuel (1972:xxxi) writes that the most fundamental question in his study is whether it is a certain category of countries, rather than a certain category of products, that tends to be victimized by unequal exchange, defined by himself as the exchange of 'a larger amount of their national labour for a smaller amount of foreign labour'. In apparent agreement with this account, and with most readers of Emmanuel's treatise in the forty years since it was published, Charles Bettelheim (1972:272) writes that 'one of the chief conclusions of this work is that increase in economic inequality between nations is rooted in "unequal exchange" ', defined as:

> the idea that on the world market the poor nations are obliged to sell the product of a relatively large number of hours of labour in order to obtain in exchange from the rich nations the product of a smaller number of hours of labour.

Nevertheless, Brolin (2006) repeatedly claims that Emmanuel's account of unequal exchange is not at all about exchanging more labour for less. If indeed most or even *all* (ibid., 347) of us have committed the same mistake of interpreting Emmanuel's argument in terms of a net transfer of labour (resulting from international wage differentials, also referred to as differences in factor costs), might not the problem be a certain lack of clarity in Emmanuel's analysis?

One reason why I have refused to concede the concept of unequal exchange to Emmanuel's contested and rather inaccessible definition is that the phenomenon of unequal exchange is much more general and inclusive than the specific structure of exchange that he identified between capitalist nations in the twentieth century. Unequal exchange in the sense of net transfers of resources has been fundamental to processes of accumulation in a wide variety of historical contexts, extending back in time at least to the earliest agrarian civilizations (see Chapters 3 and 4). Moreover, such processes of unequal exchange can be gauged in terms of several other biophysical metrics, in addition to labour. I see no reason why those of us concerned with such processes should be compelled to abandon the simple and straightforward concept of unequal exchange in favour of less appropriate and more cumbersome concepts (such as 'non-equivalent exchange'). I am thus pleased to see the recent publication of a special issue of *International Journal of Comparative Sociology* devoted to ecologically unequal exchange (Jorgenson and Clark 2009), and it is unfortunate that some researchers explicitly committed to illuminating this theme should be so constrained by scholastic quarrels as to have little of modern relevance to say on the topic. By adopting Emmanuel's conceptual framework, a student of unequal exchange will automatically become immersed in Marxist exegesis. Although Emmanuel (1972:323) explicitly rejects the 'scholasticism' of the 'endless disputes among Marxists about how the link between labour value and price of production is to be understood', his 453-page book must in itself forty years later largely be classified as scholasticism, as must a dense, 383-page treatise reiterating three centuries of economic debate and reprimanding most modern theorists of unequal exchange for not clinging to Emmanuel (Brolin 2006).

In his brief but concise 'Conclusions', Emmanuel (1972:265) writes that unequal exchange is 'one of the mechanisms whereby *value* is transferred from one group of countries to another' (emphasis added) and that 'it enables the advanced countries to begin and regularly to give new impetus to that *unevenness of development* that sets in motion all the other mechanisms of exploitation and fully explains the way that wealth is distributed'. He observes (ibid.) that economists

> have been divided into objectivists and subjectivists, but unequal exchange is denied by both parties – by one party because for them exchange is always equal in a situation of equilibrium, and by the other because for them equal exchange does not exist.

'On the basis of the classical and Marxist doctrine of labour *value*', Emmanuel advises underdeveloped countries to 'seek means to keep for themselves and prevent from leaking abroad the excess surplus *value* that they extract from their own workers' (ibid., 267; emphasis added). These quotations from his 'Introduction' and 'Conclusions' will suffice to make it abundantly clear that Emmanuel's definition of unequal exchange was initially presented as based on the idea that wage differences between countries generated international net transfers of labour 'value'. If quotations suggesting a contrary stance can be retrieved from his book, as Brolin (2006:330–331) claims to have done, it indicates a pervasive incoherence commonly identified in Marxist scholasticism. In order to account for international wage differences, Emmanuel refers to the demands of national labour movements, in part rooted in divergent cultural and historical experiences (Emmanuel 1972:126–127).

To attribute significance to the unequal exchange of embodied labour time is not necessarily to subscribe to a labour theory of value. The same applies to the unequal exchange of other productive resources such as energy or embodied land. However, to thus analytically disentangle the concept of unequal exchange from theories of value (whether of labour or land) tends to create major confusion among theorists entrenched in the habit of grounding the former in the latter. This is probably the main source of misunderstandings and disagreements on unequal exchange. Although Brolin's (2006) study is replete with such misunderstandings, which would require another round of scholasticism to disentangle, it certainly indicates voluminous reading and a commendable penchant for analytical reason. Beyond the inconsistencies, such as his contradictory treatment of Odum and Bunker, respectively, and the absence of a coherent conclusion on the actual occurrence of unequal exchange, it does provide a useful history of ideas, conveniently summarized on pp. 335–354, relating to both value and unequal exchange, from Cantillon and Quesnay through Smith, Ricardo, and Marx to Innis, Prebisch, Lewis, and Emmanuel, and finally Odum, Bunker, and Martinez-Alier. Its objection to the idea that ecologically unequal exchange is relevant to understanding uneven development is largely based on Bairoch's (1993) misguided conclusion that the 'developed West' had no need for extractive peripheries prior to 1955, a notion that I have dealt with in Chapter 3. Like Bairoch, Brolin dismisses the idea that net appropriation of natural resources had any significance for development or world-system positionality.

Carl Nordlund's (2010) study also reviews the history of the idea of ecologically unequal exchange, but as background to an empirical investigation, applying the tools of social network analysis, of actual international trade in fuels and agricultural commodities. The trade in these commodities is quantified in terms of exchange value (money prices) as well as biophysical metrics, namely energy content for fuels and embodied land (ecological footprints) for agricultural products. Nordlund's point of departure is that 'it is difficult to deny the existence of some kind of ecological unequal exchange' (ibid., 15), that the contemporary world system is 'brutally unfair in terms of resource consumption' (ibid., 18), and that 'the gaps between the haves and the have-nots' have increased with the growth in international trade (ibid., 152). His 'core question' is whether 'there is a relationship between structural positionality [in the world system] and ecological unequal exchange' (ibid., 18) and whether this is related to 'global differences in factor costs (of natural resources)' (ibid., 22), a phrasing of (ecologically) unequal exchange that is presented as more in line with Emmanuel's account than with those of more recent theorists. In Nordlund's words, while Emmanuel 'examined the national price-differentials for labour (i.e. wages), this thesis looks at price-differentials of another factor of production: land (and natural resources)' (ibid., 181). Although Nordlund's reasoning is generally clear and consistent and his methodology both innovative and sophisticated, his approach raises some conceptual issues that deserve to be discussed.

Nordlund (ibid., 178) claims that recent theorists of ecologically unequal exchange 'are not concerned with' factor costs and the 'underlying mechanisms' leading to net transfers of biophysical resources between nations, and that my own and Bunker's contributions are 'quite far away from the fundamental theoretical stanzas found in the dependency school and world-system analysis'. For obvious reasons, I cannot agree with this account. 'Factor cost differentials' is merely another way of talking about relative differences in the prices of land, labour, and capital, which for me have always been an obvious 'underlying mechanism' (but inseparable from political, cultural, and other aspects) behind unequal exchange and capital accumulation (Hornborg 1992, 1998b, 2001b). Rather than contrasting Emmanuel's concern with factor costs against others' concern with net transfers, as if they were exclusive options (Nordlund 2010:264–265), it should be obvious that they are two aspects of the same total social phenomenon of unequal exchange. Moreover, rather than attribute international wage differentials to the different cultural backgrounds of Britons, Spaniards, and French, as does Emmanuel (1972:126–127), I have consistently been concerned with their relation to world-system positionality. This is certainly more fundamental to my argument over the years than a recent 'hint' (Nordlund 2010:175).

Nordlund (2010:173–174) claims that I cannot explain 'why' the periphery 'chooses' to submit to ecologically unequal exchange benefitting the core (as if being exploited was ever a matter of choice). Apparently, my argument that the cultural evaluations of consumers are irrelevant to the physical reality of resource dissipation in economic processes (Georgescu-Roegen 1971) is interpreted by Nordlund as a dismissal of utility as a driver of consumption, but nothing could be further from the truth. Nordlund apparently confuses my dismissal of the role of utility in a thermodynamic account of economic processes with my view of its role in actual economic behaviour.

Finally, Nordlund (2010:174) suggests that a calculation that includes flows of *information* 'could very well tip the scales of the whole equation, resulting in thermodynamic unequal exchange in the very opposite direction to what is intuitively perceived'. This would run counter to the idea, expressed by Bunker, myself, and other world-system analysts, that development in core areas is tantamount to an accumulation of complexity, to the detriment of increasingly impoverished peripheries. Complexity and impoverishment *are* measures of (high versus low) information. The notion of core areas as cornucopias, exporting net flows of information to their peripheries, seems very much in line with mainstream economic theory, following an ideological tradition succinctly expressed in Rudyard Kipling's image of colonialism as the 'White Man's Burden'. We are reminded of Maurice Godelier's (1986) observation that unequal exchange tends to present itself as reciprocity, or even charity.

Although Nordlund (2010:19) appears to endorse my rejection of what Paul Ehrlich has called 'crackpot rigour', and although he momentarily expresses doubts about quantification (ibid., 264), his immersion in the technical complexities of social network analysis does suggest a kind of methodological fetishism. His preoccupation with these technical complexities finally leads him to conclusions that are very far removed from his initial conviction that the contemporary world system is 'brutally unfair'. He suggests (ibid., 264) that studies of specific commodity flows are better at identifying ecologically unequal exchange than studies of national indicators of consumption, such as ecological footprints, but I suspect that it is precisely the isolation of distinct commodity flows (fuels versus agricultural products) that in his study has more or less completely obscured the occurrence of such exchange. Thus, for example, the huge imports of fuels to the United States are never related to its huge exports of agricultural products (measured in hectares), although it should be obvious that the latter are largely made possible by the former. Instead, these agricultural exports – and flows of agricultural commodities in general – are simply interpreted as an invalidation of Nordlund's version of the hypothesis of ecologically unequal exchange (ibid., 273–277). The focus on specific commodity flows, and on the comparison of monetary and biophysical measures of these flows, may thus have obscured the total socio-ecological metabolism of which they are a part, including the conversion of imported fossil fuels into agricultural exports. The complex relation between embodied energy and embodied land, which this example highlights, will be addressed in the third and final section of this chapter. Let us just conclude that although both Brolin's (2006) and Nordlund's (2010) conclusions on ecologically unequal exchange are largely negative, their struggles have been instructive. To paraphrase Wallerstein (1995), we need to hold the tiller firm as we try to navigate between the Scylla of scholasticism and the Charybdis of methodological fetishism.

The historical contextuality of ecologically unequal exchange

From a comparative, historical perspective, it is obvious that different kinds of environmental load displacement (through trade) will accompany specific kinds of capital accumulation. We thus need to use different measures of ecologically unequal exchange for different historical and geographical contexts. What they all have in common is a concern with the factor of production referred to as *land*, a factor which, as Nordlund (2010) observes, has been largely neglected by mainstream economists over the past two centuries. Different kinds of environmental load displacement reflect the different kinds of technological infrastructures that are being accumulated, as well as the particular resource endowments offered by specific geographical circumstances. Thus, the concern with land must include not only embodied, eco-productive hectares, but also embodied energy, materials, carbon dioxide emissions, environmental degradation, water, etc. (cf. Jorgenson and Clark 2009). Different factors will be crucial bottlenecks at different times and different places. For example, nineteenth-century Europe was in great need of additional eco-productive hectares (Wilkinson 1973; Pomeranz 2000), but was more than self-sufficient in mineral energy (Bairoch 1993; Brolin 2006). Conversely, the twenty-first-century United States is in great need of imported energy, but is more than self-sufficient in agricultural land (Nordlund 2010). Against this background, it is completely logical that European colonial wars were fought over land, while contemporary American wars in the Middle East are being fought over oil. Biophysical trade balances indicate that Europe, the United States, and Japan all import significantly more materials than they export, while the converse applies to most South American countries. It is well known that per capita ecological footprints and carbon footprints are similarly skewed in favour of developed nations. Taking all these different circumstances into account is difficult but necessary, if we wish to generate a coherent understanding of ecologically unequal exchange. If we are indeed convinced that the world system is 'brutally unfair', our research questions and methodologies need to be grounded in a conceptual framework that will not be undermined by statistics that seem to invalidate a superficial, single-metric theory of unequal exchange.

Let us conclude with a final observation on the historical relation between energy and embodied land, arguably the two most likely metrics for studies of ecologically unequal exchange. Up until the Industrial Revolution, energy and land were one and the same, converging in the production of food for human labour and fodder for draft animals. For two centuries now, the age of fossil fuels has kept land requirements and energy requirements distinct from each other, making it possible for historians such as Bairoch (1993) to seriously propose that European expansion had no need for extractive peripheries (see Chapter 3). During this period, ecologically unequal exchange has not always involved net transfers of energy, nor has it always involved net transfers of embodied land, but it has always involved net transfers of *one* of these resources. As we are currently contemplating that peak oil and climate change may prompt us to turn to biofuels, we are in fact imagining a future where land requirements and energy requirements will once again coincide. Once again, it seems, it will be possible to calculate the costs of transport distances in terms of eco-productive space. What this might entail in terms of our total world-view and global social metabolism is beyond the scope of this chapter, but if we shall once again see competition over scarce land for food, fodder, fibres, and fuel, we may rest assured that the realities of ecologically unequal exchange and environmental load displacement will be recognized as very tangible conditions of human existence. In such a future, also, ecologically unequal exchange will again involve concerns with *both* energy *and* embodied land. In terms of economic theory for understanding the course of history, this would amount to the bankruptcy of both Ricardian and Marxian concepts of labour value in favour of a cosmology more akin to pre-industrial Physiocracy.

Suffice it to observe, at this point, that if the United States were to import best-practice, Brazilian ethanol (disregarding here the extent to which this ethanol is in fact subsidized by fossil fuels) to replace its current net *imports* of fossil fuels, it would require impossibly vast areas of Brazilian sugarcane, several times the agricultural area within the United States presently devoted to export production. To replace the net imports of fossil fuels to the United States in 2007 (714 Mtoe) with best-practice ethanol would have required 187 million hectares of sugarcane.[1] The current land area in Brazil devoted to sugarcane ethanol is around four million hectares. The long-term implications of the global energy shifts we shall be witnessing over the next few decades may very well lead to the conclusion that much of what we have come to know as modern or industrial technology is feasible only when it requires less local land area than the same work conducted by humans and draft animals. This has indeed been the case through two centuries of fossil fuel energy, but at the moment we have no reason to believe that this specific kind of rationality will extend beyond the fossil fuel era.

It is beginning to be possible to discern, in retrospect, the specificity of fossil fuel rationality. Access to fossil fuels – acreages of the past – in the early nineteenth century obscured our fundamental dependence on land. When acreages for horse fodder and food for workers could largely be replaced with coalmines and oil wells, present-day land became – practically as well as conceptually – a substitutable factor of production. A (re)turn to biofuels should transform economic theory in the reverse direction. We have seen that it is quite feasible to calculate the land equivalent of fossil fuels, measured in the energy content of best-practice agrofuels. In fact, *all* solar-derived energy use is analytically convertible to land area, whether human labour (hectares of food and other aspects of a labourer's ecological footprint), draft animals (hectares of animal fodder), agrofuels (hectares of energy crops), or fossil fuels (once hectares of ancient biomass, but now possible to calculate in terms of the equivalent agrofuels). This may sound like a kind of reductionism no better than the monetary reductionism of economists, but it does provide us with possibilities of rationally challenging predominant economic rationality with figures anchored in the objective constraints of the biosphere. The question we must put today remains: Was a rationale behind the transition to fossil fuels (the 'Industrial Revolution') that work performed by machines required less local land (and was thus less expensive) than the corresponding amount of work performed by humans or draft animals? And what would this imply for a post-petroleum society?

So as not to provoke undue anxiety about the prospects of a post-petroleum society, I want to add that it is a common misconception that fossil fuels are presently feeding the world. Fossil fuels in parts of the world tend to replace the work elsewhere and previously done by humans and draft animals, but there is not a Joule of fossil energy in food. Moreover, only about half of the 1.5 billion global hectares of agricultural land is currently used for basic food production for humans – the rest is devoted to fibres, lubricants, agrofuels, animal fodder, tobacco, colonial beverages such as coffee and tea, luxury fruits, etc. Of the food actually produced, perhaps as much as half is wasted or discarded. A colleague of mine has estimated that if all humans were vegans, there would be room for thirty billion of us on the planet.[2] Using organic gardening and recycled nutrients, these land areas are ample for feeding the current and even a larger global population without any fossil fuels.

7 Vital signs

How money transforms ecosystems

Human–environmental relations are being studied from a variety of perspectives in a number of disciplines. The human sciences have seen the emergence of new subfields such as environmental history, environmental anthropology, environmental sociology, environmental philosophy, and environmental economics. *Ecosemiotics*, on the other hand, should be seen as a new theoretical approach to human ecology that can be applied across several disciplines (cf. Hornborg 1996, 1999a; Kull 1998; Nöth 1998). Its basic assumptions are of a somewhat abstract and formal nature, but can be used as a framework for organizing very specific, empirical material. Its primary justification lies in its ambition to transcend Cartesian, conceptual dichotomies such as culture/nature, society/nature, mental/material, etc., that we discussed in Chapter 2. To the extent that such binary oppositions continue to obstruct holistic understandings of human agency in the biosphere, the various environmental subfields of the human sciences may have something to gain from an elementary familiarity with the ecosemiotic paradigm. One of the intriguing possibilities of this paradigm is that it can provide us with a way of fundamentally rethinking the phenomenon of money.

Briefly, we might characterize the ecosemiotic paradigm as founded on the contention that ecosystems are constituted no less by flows of signs than by flows of matter and energy. It rejects the conventional notion of nature as a primarily material phenomenon, opposed to a notion of society as primarily communicative. Rather, it views nature and society as interconnected systems, both of which are simultaneously material and communicative.

Although I wrote that ecosemiotics is a 'new' theoretical approach, this needs to be qualified. Early in the twentieth century, the zoologist Jakob von Uexküll (1864–1944) had realized how constrained our view of ecosystems had become by the obsession with quantification and materiality that dominated the natural sciences. His concern was to visualize the interaction of organisms in nature as based on their subjective, species-specific perception of each other and of their worlds. As mentioned in Chapter 2, he called such subjective worlds *Umwelten* (Uexküll 1982 [1940]). This subjectivistic brand of biology laid the foundation for the modern science of ethology, but its philosophical implications for general biology and ecology should have been far-reaching. The continued hegemony of materialism in natural science can be understood as an accommodation to the demands of an economic and technological establishment concerned with the management and control of natural systems. Uexküll's *Umweltlehre* raised more profound questions about nature than the exact measurement of its material metabolism. In echoing the animistic cosmologies of many pre-modern cultures, it has appeared romantic and of little use for the modern science of ecology. Possibly, the recent concern with biological communication (e.g. pheromones) could lead to a general reassessment of Uexküll's position. Nevertheless, this concern remains pragmatic rather than philosophical and continues to be geared to measurement and control. Whereas Uexküll and the pre-modern animists were both concerned with perceiving the natural environment as composed of sentient subjects, mainstream biology continues to convey the image of nature as an assemblage of objects.

The significance of an ecosemiotic approach becomes clearer when we consider the role of humans in ecosystems. The anthropologist, psychiatrist, and biologist Gregory Bateson (1904–1980) similarly visualized a science of living systems that focused on communication. He applied a remarkably consistent, theoretical framework to his various studies of animal behaviour (dolphins, octopuses, otters), play, alcoholism, schizophrenia, art, ritual, war, and environmental crisis. Whatever the material substrate and the particular outcome, Bateson argued that the patterns and forms of living things are generated in communicative relations between recursively engaged subjects or 'minds' (Bateson 1972, 1979). He was thus able to see cultural phenomena such as language or ritual as subsets of a much wider and more general category of communicative phenomena, which defined and coincided with life itself. This largely intuitive vision surfaced in a number of startling analogies between cultural and biological phenomena that mainstream anthropology found little use for. In later years he explicitly tried to apply these insights to environmental concerns by approaching ecological crisis as a 'pathology' of epistemology and communication. Such normative and functionalist arguments made him and his followers (e.g. Rappaport 1968, 1979) easy targets for criticism within anthropology (e.g. Friedman 1979), but I believe that this has been an unfortunate case of throwing the baby out with the bathwater. As Bateson and Rappaport envisaged, we have many insights to gain from viewing ecological crisis as a problem of communication.

In this chapter I shall address three questions that I believe to be of fundamental importance to ecosemiotics:

1 To what extent can ecosystems be seen as semiotic (sign-mediated) phenomena?
2 To what extent can those semiotic flows and processes that are organized by humans be seen as constitutive of ecosystems?
3 What are the different kinds of human sign systems that take part in the constitution of ecological processes, and which generalizations can we make about their respective roles in transforming ecosystems?

The first of these questions challenges the conventional dichotomy of the communicative or mental (cf. Bateson's notion of 'mind') and the material. The second challenges the dichotomy of culture and nature. The third, finally, mobilizes ecosemiotic perspectives in an attempt to provide a general account of ecological crisis. It suggests, in fact, that such perspectives should be crucial ingredients in the currently ongoing deliberations on global sustainability.

Although the issues raised are of general and global importance, I have chosen to ground this discussion in the ethnography and historical ecology of Amazonia. Amazonia is the region of the world the ethnography of which I am most familiar with (Hornborg 1988, 1993, 1998a, 2005; Hornborg and Hill 2011), it is also a bioregion conventionally perceived as pristine wilderness yet inhabited for millennia by human populations that have actively transformed it. Also, an increasing volume of modern ethnography has been documenting in subtle detail the indigenous cosmology and practice of human–environmental relations in the region. Needless to say, this chapter permits me only to sketch the barest outline of such a comparative study, the main purpose of which is to convey the wide spectrum of possible applications of an ecosemiotic perspective. Due to the wealth of relevant material, only a minute sample of what the ethnographers of Amazonia have reported on human–environmental relations in the area can be mentioned in this context.

In posing and addressing questions such as those above, I hope that we shall be able to organize our thinking about human–environmental relations in new ways. In particular, an ecosemiotic approach may provide a conceptual framework for transcending Cartesian dichotomies such as mental/material or culture/nature without abandoning the analytical rigour that remains Descartes' more essential legacy.

1 To what extent can ecosystems be seen as semiotic (sign-mediated) phenomena?

As Uexküll and Bateson have both shown in different ways, the material interactions of organisms in ecosystems presuppose their exchange and interpretation of signs. This point requires no elaborate argument, but a single example may be appropriate. The harpy eagle (*Harpia harpyja*) of the Amazon rainforest is equipped not only with a very keen sense of sight but also with a propensity to respond to those particular kinds of movement in the foliage that result from the activities of monkeys. Were it not for the eagle's capacity to interpret such very specific sense data, it would not maintain its position at the apex of the Amazonian food chain. Conversely, monkeys have responded phylogenetically by becoming increasingly attentive to signs of eagles. This can be generalized for the entire rainforest ecosystem. In a myriad similar ways, each organism and species exists by virtue of its capacity to perceive and interpret the world around it. An ecosystem is not a machine, where the various components mindlessly fulfil their functions as a reflection of the external mind of the engineer. Ecosystems are incredibly complex articulations of innumerable, sentient subjects, engaging each other through the lenses of their own subjective worlds. Intriguingly, as we shall see, such an interpretation of ecology, here offered in an academic context, harmonizes very well with traditional, indigenous ecocosmologies (Croll and Parkin 1992; Århem 1996) in Amazonia. Ecosemiotics thus does not merely provide a vantage point for understanding these cosmologies in theoretical terms, but actually also for validating them.

2 To what extent can those semiotic flows and processes that are organized by humans be seen as constitutive of ecosystems?

As we discussed at length in Chapter 2, there is a tradition in European thought of maintaining a conceptual boundary between Nature and Culture. To be sure, as illustrated in Chapters 3 and 4, the impact of human activity on the natural environment has been apparent throughout Western history. However, rather than acknowledging the pervasive interfusion of humans and non-human nature, European cosmology has responded either by ignoring the anthropogenic aspects of landscapes or by reserving its notion of nature for distinct but dwindling, geographical areas perceived as uncontaminated by human activity (cf. Ellen 1996). This is well exemplified by a quote from Marston Bates' *Where Winter Never Comes*, introducing a popular volume on *The Amazon* published by Time-Life Books as part of its series on *The World's Wild Places*: 'It is a humbling experience, and surely a healthy one, to enter a landscape that man has not been able to alter, to dominate, to twist to his own purposes' (Sterling 1973: 18).

In recent years, there has been a growing recognition of the need to transcend such dichotomies of Nature versus Culture, for instance in our approach to landscapes, human bodies, and new technologies (see Chapter 2).

Both human organisms and their environments are partly biophysical and partly symbolic in derivation. Nature and Culture are not mutually exclusive domains or essences, but sedimentations of semiotic processes at different levels of integration. Even biochemical processes have a semiotic dimension (Hoffmeyer 1996), not to mention the various visual, auditory, olfactory, and tactile communication systems of human and non-human organisms. It nevertheless remains justified to analytically distinguish cultural phenomena as those which rely on that specific category of signs that we call *symbols*, defined by Peirce (1931–1958) as signs that relate only by convention to the objects to which they refer. Cultural or symbolic phenomena may thus qualify as more or less uniquely human, yet in this context should be recognized as recent additions to the more general semiotics of ecosystems.

We may note in passing how difficult it is to extricate ourselves from dichotomous patterns of thinking that are evoked by polarizing vocabularies such as nature/culture, material/communicative, or biophysical/symbolic. Such distinctions need not be abandoned, however, as long as we acknowledge them as referring merely to aspects or dimensions of living systems rather than to concrete segments of reality. Seen as aspects of integrated wholes rather than entities in their own right, the opposite poles of such analytical dichotomies should not be treated as standing in some kind of causal relationship to each other. An analytical dualism, in other words, need not imply an ontological dualism.

Amazonia provides us with remarkable evidence for the extent to which human, cultural behaviour can be constitutive of ecosystems. I am not referring here to the widely publicized deforestation of significant parts of the region in recent decades, but to the discovery that the composition of the presumably pristine rainforests themselves has been influenced by human activity over the centuries. It has been suggested that at least 12 per cent of the *terra firme* forests of Brazilian Amazonia may be anthropogenic, i.e. 'of a biocultural origin that would not have existed without past human interference' (Balée 1989, 1993: 231). Detailed, ethnobotanical studies of the agroforestry practices of the Kayapó (Posey 1983, 1985) and the Ka'apor (Balée 1993) of Brazil reveal that human activity has had a tendency to enhance rather than diminish biodiversity. In stark contrast to modern monocultures of, for example, eucalyptus or soya beans, indigenous agroforestry complexes tend to increase the number of species per hectare. Whether or not this is the result of conscious intention is a matter of contention (Balée 1993), with obvious implications for discussions of the nature of 'traditional ecological knowledge', yet does not really affect the general conclusion here, namely that the cultural predilections of human beings leave their marks even in the most natural of environments.

A fundamental challenge for ecosemiotics is to develop a framework for representing the interpenetration of culture and nature. Although there seems to be a widespread consensus that the dichotomy needs to be transcended, most contributions that make this claim can be more or less readily assigned to either of C. P. Snow's 'two cultures' (the humanities versus natural science; cf. Ingerson 1994). On one hand, there are studies that in various ways are concerned with the human experience and perception of the environment (e.g. Croll and Parkin 1992; Hirsch and O'Hanlon 1995; Descola and Pálsson 1996; Ellen and Fukui 1996; Ingold 2000). On the other hand, there are studies that tend to privilege quantitative data on the measurable, biophysical parameters of human–environmental relations (e.g. Moran 1990, 1993; Crumley 1994; Sponsel 1995; Bates and Lees 1996; Kormondy and Brown 1998; Wilson 1999). Over the years, although several anthropologists have demonstrated proficiency in both discourses (e.g. Rappaport 1968; Ellen 1982; Ingold 1986; Descola 1994), very rarely has there been a successful, theoretical integration of humanistic and biophysical dimensions in the study of human ecology. In the present discussion I can only in a very rudimentary manner suggest some theoretical avenues for striking such a balance between subjectivism and objectivism. Let us begin by asking the kinds of fundamental questions that emerge from an ecosemiotic perspective.

3 What are the different kinds of human sign systems that take part in the constitution of ecological processes, and which generalizations can we make about their respective roles in transforming ecosystems?

I shall distinguish between three kinds of sign systems: *sensory*, *linguistic*, and *economic*. Each of these semiotic levels is a prerequisite for the next, since linguistic signs must be mediated by sense organs and economic signs by cultural categories pertaining to exchange (e.g. money, price, commodity, wage). On the other hand, each level has had a tendency to progressively detach itself from the logically and phylogenetically prior one, disembedding discourse from experience and economy from culture.

Sensory signs

To begin with, of course, humans are equipped to transmit and receive sensory (visual, auditory, tactile, olfactory, taste) signals like other animals. Such pre- or extra-linguistic sign systems are intrinsically difficult to theorize about, primarily because theory itself is founded in language. Theory can here rarely do more than evoke what remains an infinitely subtle, elusive, and largely unconscious level of human experience. Yet the approach of phenomenology, with its notions of dwelling and 'being-in-the-world', has been quite successful in helping us acknowledge its importance. This in itself has amounted to a powerful counterpoint to materialist science and Cartesian objectification. Significantly, phenomenological approaches in environmental philosophy (Evernden 1985) and anthropology (Ingold 2000) tend to emphasize the fundamental, human inclination to experience the natural environment as composed of subjects (cf. also Bird-David 1993, 1999).

Throughout millennia of foraging and subsistence horticulture in Amazonia, a major part of the interaction between human and non-human organisms has been mediated by a myriad sensations of the eye, ear, nose, tongue, and skin, only a fraction of which have been reflected upon and assigned linguistic categories. Such sensory sign flows are what constitute the human embeddedness in the world evoked by phenomenologists and so-called practice theorists. To the extent that people mimetically reproduce and share conventional patterns of emitting and responding to such sensory signals, these patterns are cultural. To the extent that they represent embodied, practical skills of coping in the rainforest environment, they can be classified as a non-linguistic component of traditional ecological knowledge. This sensory level of human–environmental relations includes modes of interpreting non-human life forms as well as modes of communicating with them. Amahuaca hunters in Peru, for example, know not only how to glean information from sounds, movements, scents, excrements, tooth marks, tracks, bits of fruit, displaced leaves, broken twigs, etc., but also how to disguise their own colour and scent and to imitate animal cries and try to get the animals to respond (Carneiro 1974 [1970]:126–127). Hunters among the Achuar of Ecuador similarly use all their senses in interpreting the characteristic signs of different species and are able to 'do a perfect imitation of the distress calls of young or of a female in heat of any species to draw the parents or males within range of the blowgun' (Descola 1994:237).

Though seemingly ephemeral and largely uncodified in language, sensory sign systems are nonetheless potent ingredients in human–environmental relations, capable of inscribing themselves in the landscape. The relation between such local, cultural experience and natural surroundings is clearly co-evolutionary, or recursive, much as any other relation in an ecosystem. Whether deliberately or not, the dietary and other cultural preferences of past generations of Amazonian Indians such as the Ka'apor have left a tangible record in the form, for instance, of old fallows, with a much higher incidence of food species (Balée 1993:245, 1995:106). Referring to Carole Crumley's definition of landscape as 'the material manifestation of the relation between humans and the environment', Balée (1995:106) concludes that 'old fallows constitute landscapes par excellence'. The Huaorani of the Upper Marañon similarly tend to encourage the occurrence of the *ungurahua* palm (*Jessenia bataua*) and other useful species, while more consciously cultivating the peach palm (*Bactris gasipaes*) and sweet manioc (Rival 1996:238–241). Rival shows that these species are assigned different symbolic values and associated with different kinds of social relations. The peach palms and their fruit are perceived as gifts from deceased relatives and appropriate food for celebrating within the endogamous group, while manioc is used to entertain visitors and potential allies. It is not difficult to imagine the role of such sentiments in generating specific patterns of distribution for different plant species. Although much of this crop symbolism is evidently codified in explicit preferences, it is apparent that the sentiments thus expressed represent a more elusive, sensory level of experience that is transmitted largely through mimetic practice rather than words. No less than language, social practice is a cultural process that conditions human beings to respond in specific ways to particular signs. Suffice it to say that sensory, non-linguistic signs do play an important role in traditional, human–environmental relations, and that, for millennia, they have been active ingredients in the human transformation of ecosystems.

Linguistic signs

The paradigmatic human sign system, of course, is language. Linguistic sign systems are generally regarded as the central medium of culture, and it is no coincidence that anthropological theories of culture to such a large extent have been inspired by linguistics. Whether analysing systems of ethnobiological classification (Fowler 1977; Berlin 1992) or metaphorizations of human–environmental relations (Lévi-Strauss 1969b, 1973, 1978; Bird-David 1993; Descola 1994; Århem 1996; Viveiros de Castro 1999), anthropologists have focused on linguistic data. Cultural categories and metaphors are likely to influence the way humans engage their natural surroundings, but for anthropologists of a phenomenological persuasion, such linguistic codifications are secondary to practical knowledge, or skill (cf. Ingold 1992, 2000). The relative significance of the two levels probably varies between different cultural contexts, with modernization generally entailing increasing emphasis on language. In any case, linguistic signs have always been the main source of cultural information for anthropologists, and it is difficult to envisage a systematic research methodology for eliciting cultural information that would not have to use language as a medium.

Human language is definitely an active ingredient in ecosystems (cf. Rappaport 1993:156). Balée (1995:100–101) suggests that one of the effects of post-conquest depopulation on the Guajá Indians of Brazil was a significant reduction of plant vocabulary, and that this was associated with their abandonment of a sedentary and agricultural existence. (He nevertheless finds in the modern, impoverished Guajá language compelling evidence for prior horticulture.) The number of generic plant names among the linguistically related and still agricultural Ka'apor is about 479, while in Guajá (for the same flora) the number is 353. It is evident that language may serve as a repository of ecological knowledge, if only as a mnemonic device providing cues for the activation of a much larger body of practical knowledge acquired through practice. Language can thus be seen as a codification of human practice, including practical engagement with non-human organisms. In serving as a kind of systemic memory that is indeterminately yet significantly related to performance, linguistic codes are in certain respects analogous to genetic codes (Hornborg 1988). This applies not only to ethnobiological taxonomies such as those investigated by Balée, but also to the various metaphorizations of human–environmental relations that have been reported from indigenous peoples throughout Amazonia.

The social life of nature

Metaphorical representations of this kind often invoke familiar aspects of social life to serve as models for human interaction with non-human nature. Descola (1992) refers to such projections of meaning from society to nature as 'animism', as opposed to the inverse projection from nature to society that we know as 'totemism' (cf. Lévi-Strauss 1966). Common to both is the interpenetration or perhaps amalgamation of categories and sentiments pertaining to social and human–environmental relations, respectively. Ingold (1996) argues that these two kinds of relations are not distinguishable from each other in hunting-and-gathering societies. From the perspective of an outside observer, however, it is striking how widespread is the drift of attitudes and sentiments from social to human–environmental relations in traditional societies (cf. Bird-David 1993; Descola 1994). No doubt there is also a widespread transfer of meanings in the opposite direction, suggesting a dialectic between animism and totemism.[1]

Among the Achuar, Descola (1994:327) discovers that 'the distinction between hunting and gardening is nurtured by an opposition between two types of sociability: the women's consanguine mothering of cultivated plants and the affinal charming of game practiced by the men'. Skills in these activities hinge on the capacity to communicate with other life forms through magical songs called *anent*, and Descola (ibid., 262) notes that hunting *anent* 'have a cajoling, wheedling tone that is not found in gardening *anent*, the garden being a world of kindred beings devoid of the touchy feelings that the man must take care not to ruffle in dealing with allies'. In line with what has been previously said, it is obvious that the linguistic form of these songs is merely the skeleton of a profoundly embodied mode of communication. The word *anent* literally designates 'words that come from the heart' (ibid., 198). The 'sung' mode, says Descola (ibid., 200), is used 'to overdetermine normal language in those circumstances where the latter is not an adequate vehicle, that is when the words must reach the heart of a spatially or ontologically remote target'. In the Achuar universe, humans, plants, animals, and even meteors can be seen as persons with individual souls (ibid., 93), and Descola asks why, given such a universe, the Achuar should forego the normal means of acting upon it. 'When one postulates that cultivated plants are beings with souls, it is obviously normal to attempt to keep up harmonious relations with them, using for this magical songs (*anent*) which are employed to the same end in human relations' (ibid., 214).

It is evident that the essence of those practices that modern people refer to as magic or ritual is the imperative to communicate with non-human nature that stems logically from viewing the world as composed of sentient subjects. This attitude may be extended beyond the domain of living things, as when the Eastern Timbira of Brazil ceremonially induce the Moon (*Puduvri*) to keep the maize plants free of animal parasites (Nimuendajú 1974 [1946]:117). If astronomical bodies are capable of signalling to humans – as indeed the Pleiades (*krot*) to the Eastern Timbira signal the rainy season and the need to start clearing for a garden (ibid., 116) – it is only logical that they should be involved in the total, communicative community of which the Amazonian Indian perceives himself to be a part. It should nevertheless be justified to distinguish between those human signs that actually do have a sentient, non-human receiver (e.g. in the form of a game animal) and those that only ostensibly do so (e.g. exhortations to astronomical bodies). This is not to say that fictitious communication is necessarily less efficient in producing the desired result than actual communication, as it may serve the important purpose of calibrating an uncertain subject to its task. Tambiah (1968:202) writes that it is possible to argue that 'all ritual, whatever the idiom, is addressed to the human participants and uses a technique which attempts to re-structure and integrate the minds and emotions of the actors'. From the perspective of an observer, however, there is clearly a difference between actually communicating with non-human subjects, on the one hand, and individually or collectively summoning one's own powers of concentration, on the other.

In the eye of the beholder

Indigenous groups in Amazonia have developed their own cultural lenses for interpreting their natural environment, but there are some fundamental themes that seem to be pervasive throughout most of this vast region. One such theme is a keen awareness of the notion of perspective, i.e. the recognition that any particular perception of the world is contingent on the vantage point of the beholder. In acknowledging that the relation between sign and object hinges on the interpretant, this basic wisdom reveals an indigenous, Amazonian concern with semiotics. An early observation along these lines is Weiss's remark about the cosmology of the Campa:

> It is a world of relative semblances, where different kinds of beings see the same things differently; thus human eyes can normally see good spirits only in the form of lightning flashes or birds whereas they see themselves in their true human form, and similarly in the eyes of jaguars human beings look like peccaries to be hunted.
>
> (Weiss 1974 [1972]:264)

A number of Amazonian ethnographies confirm this pervasive link between what Århem (1991) and Viveiros de Castro (1998) call *perspectivism* and an understanding of the world as a continuous food chain. To the Makuna of south-eastern Colombia, for instance, any category of beings will classify all others as either predators or prey, in shamanic language referred to respectively as *yai* (jaguar) and *wai* (fish) (Århem 1996). As among traditional hunting peoples in many parts of the world, the relation between the Makuna hunter and his prey is construed as a reciprocal exchange between human and animal communities, modelled on the relation between affines. Through a continuous recycling of souls or generic vitality, predation – whether in the form of humans consuming animals or spirits consuming humans – is thus also procreation. This is linked to a widespread, cosmological theme that represents affinal relations and predation as cognate forms of symbolic exchange (cf. Descola 1992; Overing 1993). This theme has inspired a number of anthropological studies of what Viveiros de Castro (1996:190) calls the symbolic economy of alterity, linking native Amazonian sociologies and ecologies into a common, interpretative framework for understanding as seemingly disparate, cultural phenomena as war, cannibalism, hunting, shamanism, and funerary rites (cf. Viveiros de Castro 1992).

A crucial element of such animistic cosmologies of Amazonia is the conviction that all beings are fundamentally humans or *persons*, as expressed, for instance, in the Makuna notion of *masa* (Århem 1996:200). Viveiros de

Castro (1999:4) writes that to say 'that animals and spirits are people, is to say that they are persons, and to personify them is to attribute to non-humans the capacities of conscious intentionality and social agency which define the position of the subject'. Whereas Europeans have had a tendency to think of humans as composed of a cultural surface that disguises their animal essence, Amazonians see it the other way around: the surface appearance of animals disguises their human aspect (ibid., 3). For Amazonian Indians, to know is to personify rather than to objectify, and an object is merely 'an incompletely interpreted subject' (ibid., 6, 8). Against this background, shamanism can be understood as the highly valued capacity of certain humans to adopt the perspective of non-human subjects (cf. Århem 1991).

Viveiros de Castro (1999:10) argues that Amazonian perspectivism is not tantamount to relativism: all beings see the world in the same way, i.e. in terms of a single set of fundamental categories; what changes is the world that they see. Drawing on an observation by Renard-Casevitz on the Matsiguenga of Peru, he suggests that words that to us denote objective, self-contained entities (such as fish, snake, hammock, or beer) tend to be used by Amazonians as relational pointers in the same way as kinship terms. Thus, 'what to us is blood, is maize beer to the jaguar; what to us is soaking manioc, the souls of the dead see as rotting corpse; what we see as a muddy waterhole, the tapirs see as a great ceremonial house' (ibid., 10; cf. Århem 1991:119–120). In these cosmologies, it seems, to be a fish is not an intrinsic property but a condition established 'only by virtue of someone else whose fish it is' (Viveiros de Castro 1999:11). I interpret this to mean that the category fish should not be confused with the actual set of species to which humans feel that it refers, but should rather be understood as an interspecific, abstract sign for food, of which these species, from the point of view of humans, are merely an instantiation. There is an underlying unity of soul in all beings which causes them to see the same things everywhere; the difference in how they perceive other beings (as belonging to one or the other among their common set of categories) is given in the specificity of their bodies (ibid., 13). What is remarkable about these cosmologies, from a modern vantage point, is the extent to which Amazonians have acknowledged the limitations of their own, human powers of perception, and the empathy with which they have imagined other species' ways of viewing the world. Yet, all through this radical decentring of human experience, their basic code of linguistic categories appears never to be cast in doubt.

Food taboos and the attempted mechanization of culture

In considering various, linguistically mediated aspects of human–environmental relations in Amazonia, we should finally say something about dietary prohibitions, i.e. food taboos. Twenty years ago, this topic became the focus of a more general debate that opposed materialist and mentalist interpretations of human–environmental relations in Amazonia and elsewhere. On one hand, the approach of cultural ecology characteristically viewed dietary prohibitions as functional adaptations to the constraints of the natural environment, e.g. as pragmatic responses to the imperatives of resource management (Ross 1978). On the other hand, several anthropologists instead argued that the rationale for such prohibitions should be sought not in nature but in culture itself, whether approached from a symbolic, structural, or psychological perspective (Kensinger and Kracke 1981). Some of the latter emphasized, however, that there need be no real contradiction between the two approaches (e.g. Menget 1981:6). Even the semiotician Urban (1981), who initially declares himself to be a 'mentalist', concedes that 'food taboos have ecological consequences, and so can be profitably analysed within the framework of an ecosystem perspective' (ibid., 88, note 1). Tuzin (1981:190) concluded that the issue 'is not whether we see the boundary between mentalism and materialism to be friendly or hostile, but that we see the boundary at all'. As long as we do, he continues, 'there will be no unified theory of food taboos … or culture in general'.

The main point of these studies seems rather to have been to show that the semiotics of food taboos are more than automatic reflections of the exigencies of the environment, and that the logic of cultural meanings has an autonomy and a specificity that accords with a view of human populations as active and idiosyncratic subjects. Seen in this light, it seems that these anthropologists were provoked by the ecologists to do for human populations precisely what Uexküll and his followers have sought to do for non-human species: to grant them the status of subjects. The paradigm of cultural ecology, not to mention the cultural materialism of Harris (1979), tends to extend the denial of subjective agency from mainstream biology's mechanistic view of ecosystems into human society and culture. An ecosemiotic perspective, on the contrary, would grant human meaning systems the same measure of idiosyncrasy as the *Umwelt* of any other species, and perceive ecosystems as the stochastic outcome of the coexistence of a multitude of such subjectivities. Co-evolution is clearly a better word for these processes than the cultural ecologists' notion of adaptation, which conjures the image of a one-way learning process, geared to a static environment and leaving no room for creative, idiosyncratic innovation. Rather than amount to a distinction between the autonomous subjectivity of culture and the mechanical pragmatics of nature, humanist arguments can be accommodated within a more sensitive, communicative theory of life. As we have seen, such a theory might well have much in common with those to which many Amazonian Indians have traditionally subscribed.

The evidence for cultural idiosyncrasy is quite obvious as far as the semiotic logic of food avoidances is concerned (cf. Leach 1964; Tambiah 1969; Sahlins 1976). Among the Shokleng of Brazil, food taboos 'far transcend their purely ecological functions' by signalling social status and thus mediating social relations, a phenomenon which can be recognized as belonging to the widespread phenomenon of totemism (Urban 1981). From a psychoanalytic perspective, Kracke (1981) argues that food avoidances among the Kagwahiv can be understood as a symbolic language for articulating and resolving repressed, personal conflicts. The Kagwahiv thus 'use nature as a rich source of metaphor for depicting emotional states and intimate relationships' (ibid., 133). Kracke shows how the domestication of non-human nature with human qualities rebounds into society by offering a code with which to express social relations:

> The plaintive call of the jacamin, the wail of the jogo-jogo, and the more raucous cry of the toucan are identified with a baby's crying, while the tapir is a sexual competitor, the paca self-indulgent, and the agouti and tinamou lazy.
>
> (ibid., 130)

Human qualities are thus projected onto animals prior to their serving as ordering principles for society, much as was observed above about the dialectic between animism and totemism.

Such recursive processes of meaning transfer are eminently suitable for semiotic analysis. Urban (1981:86) observes that the purported consequences of transgressing a dietary restriction reveal 'ethnotheoretic notions that can be conveyed only through language', and that 'turn out to involve an "iconic" (or what used to be called "sympathetic") connection between the species and supposed consequence'. For instance, the Shokleng claim that eating paca or agouti meat would cause the teeth of a child to grow too rapidly, causing toothache. Similarly, the Sanumá on the border between Brazil and Venezuela say that parents eating snake meat may cause their children to have diarrhoea, since snakes have liquid excrement, and that eating sloth meat may cause them to develop a twisted wrist (Taylor 1981:43–44). Kagwahiv maintain that the infant child of a man who kills a curassow (a red-beaked bird) may develop inflammations of the mouth and lips (Kracke 1981:114). Referring to Leach (1964) and Tambiah (1969), Kracke (1981:110) suggests that food prohibitions can be accounted for in terms of the different species' metonymical or metaphorical proximity to humans (pets are metonymically close, while monkeys are metaphorically so). Following Lévi-Strauss (1966), Descola (1992: 114; 1994: 211) observes that some animal or plant species are particularly well suited to the role of symbolic signifier because of distinctive, visible features that suggest invisible properties. Thus, dietary prohibitions recognized by the Achuar at the time of planting their gardens function

> as a sign pointing to one of the three categories of attributes detrimental to plants' harmonious growth: things that rot, signified by the *kanka* fish, the *muntish* grub, and by digestion in general; things that burn, signified by peppers and meat exposed to direct contact with fire; things that are slender, signified by monkeys swinging on flexible branches.

Such semiotic transformations are evident not only in food prohibitions, but throughout all the various aspects of indigenous cosmology. For the Campa, whatever is 'excessively thin', has the 'drab colour of decay', or 'presents a false appearance' is a demon; thus, *shiénti* (adult ant lions), *tsiisanti* (drab-coloured hummingbirds), and *shiinti* and *tsináro* (leaf-like katydids) are all demons (Weiss 1974 [1972]:262). With regard to the last of these categories, we can observe in passing that the indigenous concern with deception – as in the widespread, mythological theme of the Trickster – is a matter deserving rigorous semiotic analysis. Deliberate deception is obviously not the same thing as when the human essence of an animal is not apparent to the human observer, due to his or her limited powers of perception, but both are profoundly semiotic concerns.

The semiotic logic underlying indigenous Amazonian sentiments regarding plant and animal species thus cannot be reduced to objective, utilitarian principles that somehow transcend the vagaries of sensory experience and symbolic classification. In fact, the claim of some Western observers to have access to these transcendental principles – whether adaptation, optimization, or even sustainability – can in itself be taken as an expression of a particular symbolic scheme (cf. Sahlins 1976; Gudeman 1986). However, to say that indigenous cosmologies are not immediately adaptive in a mechanical sense is not to deny that their fundamental, relational modality of human–environmental calibration, in all its attentiveness to the ecological Other, seems singularly attuned to the vital task of communication. This mode of ecological calibration, as we have seen, is founded on the continuous interweaving of direct, sensory experience and local, oral communication.

Linguistic signs, however, may also be subjected to different degrees of social reification and decontextualization. The invention of writing facilitated the detachment of words from local, socio-ecological negotiations and their use as top-down signalling systems in expansive, hierarchical social systems. Textualized knowledge systems with universalist aspirations are crucial vehicles for implementing modernist monocultures in Amazonia as elsewhere. The fetishized vocabularies of economics and agricultural science are not designed to stay attentive to the integrity of specific, local ecosystems, but to pursue their own, abstract designs on an increasingly standardized, global landscape. This transformation in the socio-ecological use of language has been intimately geared to the logic of economic signs.

Economic signs

A third kind of ecosemiotic flows are the movements of artefacts, people, resources, and exchange values that comprise the subject matter of economics, economic history, and economic anthropology. As we have already noted, such flows presuppose specific, cultural and linguistic understandings of exchange, including notions about reciprocity and its appropriate, institutional frameworks. However, as investments of human labour and natural resources congeal into material exchange values, the circulation of such artefacts acquires a logic of its own that transcends the face-to-face reciprocities of oral communication and local value systems (cf. Munn 1986; Santos-Granero 2009). Such flows have long been an influential factor in shaping human–environmental relations in Amazonia, but their ecological repercussions have recently seen a drastic increase in magnitude.

As we concluded the previous section by discussing the semiotics of food, it may be appropriate to begin by recalling, with Kelekna (1981:178), that Lévi-Strauss (1969a) juxtaposed the exchange of words, objects, and women as constitutive of traditional societies, and that 'food occupies an unusually prominent place in the category of "objects" exchanged between individuals and groups'. Classical monographs discussing the social significance of food exchange among Amazonians include Chagnon (1968) on the Yanomamö, Rivière (1969) on the Trio, and Siskind (1973) on the Sharanahua. A corollary of what was previously said about food avoidances is that foods that are positively enjoined (and enjoyed) will also be so largely by virtue of some symbolic scheme of classification. Kelekna (1981:184) observes that a person through the ritual consumption of a prescribed item may attempt to 'appropriate the power manifested by a plant, a fish, a bird, an animal', a phenomenon that extends even to cannibalism (Kracke 1981:104; Viveiros de Castro 1992). Certain crops and their derivatives may also be assigned special, ritual significance, as has been the case with maize and maize beer in many areas of South America (Morris 1979).

Foodstuffs not only serve as lubricants for local, social relations, but frequently also become important items of trade. Distantly imported foodstuffs and other consumption goods have always been an important source of elite identity in stratified societies. Various geographical constellations of local, cultural demand and sources of the coveted products have generated regional systems of exchange linking various parts of Amazonia with each other and with adjoining regions such as the Andean highlands (Gade 1972; Lathrap 1973, 1974; Roth 1974 [1924/1929]; Oberem 1974 [1967]; Camino 1977; Myers 1983; Whitehead 1993, 1994; Arvelo-Jiménez and Biord 1994; Hornborg 2000, 2005; Eriksen 2011). The widespread distribution of prehistoric art styles may also be indicative of long-distance exchange, e.g. of ritual paraphernalia (Lathrap 1974; Roosevelt 1994:6). Drawing on archaeological evidence from pre-conquest Amazonian chiefdoms, Roosevelt (1993:260) writes that strings of 'disc beads, usually of shell, were widely used as a medium of exchange, and semi-precious stone ornaments, such as greenstones [jade], were part of a system of elite gift-giving'. Trade goods mentioned by Whitehead (1993:295, 1994:38) include worked jade from the lower Amazon exchanged for goldwork from the lower Orinoco and circum-Roraima region, poisons, metals, and jewels from the Guayana highlands, and gold objects 'traded out of the Vaupés-Rio Negro region by the Manoas and circulated eastwards, as far north as the Caribbean'. Even if much of the evidence for indigenous, long-distance trade derives from post-conquest historical sources, such extensive systems of exchange – and thus ultimately the global, cultural logic of consumption – have undoubtedly been important factors in determining patterns of land use in Amazonia since prehistoric times. As the archaeological evidence is generally restricted to durable trade goods such as stone, metal, and shell, it is important to remember that the greatest volumes of trade may have been in perishable, organic materials such as herbs, feathers, and animals.

Long-distance exchange in valuables may significantly affect local land use both at their source and their point of consumption by encouraging intensified production and providing catalysts for stratification. In premodern Amazonian societies, long assumed to have been intrinsically egalitarian (cf. Clastres 1977), it may have provided one of few possibilities for gaining political power through the control of scarce resources (alongside, for example, control of fertile river bottom lands, or of women, or sons-in-law, or ceremonial knowledge; cf. Kracke 1993). Archaeological evidence for prehistoric stratification and agricultural intensification is accumulating throughout Amazonia, with particularly striking settlement sizes and earthworks reported from Marajó Island at the mouth of the Amazon, the Santarém area on the lower Amazon, and lowland areas of Bolivia and Ecuador (Roosevelt 1993, 1994:8). The ridged fields and causeways of the Mojos area of Bolivia cover thousands of hectares and are clearly visible from the air. The earliest historical sources confirm high population densities, stratification, and a lively trade along the major rivers prior to European conquest (Porro 1994). Early documents emphasize the intense exploitation of floral and faunal resources and the high levels of agricultural productivity of these societies (Whitehead 1994:36). Whitehead (ibid.) notes that the

> fact that some Amerindian economies were geared to producing such food surpluses in antiquity is also demonstrated by the existence of indigenous markets and exchange systems in fish meal and manioc flour, as well as the large-scale 'ranching' of turtles and iguanas.

He suggests, furthermore, that stratified trading polities (exemplified by the Lokono of Guayana) would have thrived as much in the interfluvial areas as in the floodplains, as such locations would have afforded control over important trade routes (Whitehead 1993:296, 1994:36).

In the sixteenth century, the main rivers of Amazonia were evidently dominated by complex, densely populated societies practicing intensive cultivation of seed crops such as maize (Roosevelt 1993). Some archaeological sites suggest populations 'in the tens of thousands at least' (ibid., 274). In line with a general argument that goes back to Lathrap (1968), Roosevelt (1993) suggests that the subsistence practices of most contemporary Amazonians (swidden manioc horticulture, fishing, and hunting) over much of the area represents a post-conquest regression to patterns that prevailed before the emergence of these complex polities, beginning in the first millennium BC. This process of devolution was precipitated by the traumatic, demographic collapse that occurred between the mid-sixteenth and mid-seventeenth centuries as a result of the introduction of European microbes, yet largely prior to actual, European settlement (Porro 1994; Whitehead 1993:289). These microbes rapidly diffused along the vital arteries of indigenous exchange, devastating the social fabric of Amazonia. The very existence of an established, economic communication network integrating the region was thus conducive to biological contamination, depopulation, and social crisis. There were nevertheless also indigenous groups who were able, at least temporarily, to take advantage of new opportunities for trade following the arrival of the Europeans (Whitehead 1993:287, 297, 1994:41).

Largely depopulated, Amazonia in the eighteenth and nineteenth centuries reverted to the semblance of a pristine wilderness, but the growing colonial population engaged in various kinds of natural resource extraction initiated new processes of very tangible changes in land use. Whitehead (1993:293) writes that the colonial process 'has not been limited to the social sphere but also extended to the biological, directly resulting in the degradation or destruction of aboriginal flora and fauna'. Echoing Crosby's (1986) observations on ecological imperialism, he mentions 'the rapid way in which wild cattle, pigs, and goats may have directly degraded the grassland-savannas making them less favourable to the indigenous species' (Whitehead 1994:37). The more recent and ongoing incorporation of Amazonia in a global market economy has accelerated environmental destruction (Brondízio 2006). The local linkages between political economy and environmental destruction (such as deforestation) have been phrased in terms of political ecology (e.g. Painter and Durham 1995), and perspectives from dependency theory and world-system analysis have been applied to illuminate the global logic of unequal exchange that continues to impoverish the extractive economies of Amazonia (Bunker 1985). The rainforest ecosystem is today threatened not only by commercial forestry, ranching, and agriculture, but also by the extraction of subterranean resources such as oil.

From an ecosemiotic perspective, it is obvious that modern money and commodities are signs capable of radically transforming, if not dismantling, Amazonian ecosystems. The capacity for destruction that seems to be inherent in modern, general-purpose currency is an effect of the peculiarly vacuous, semiotic properties of money (Hornborg 1999a). Briefly, this semiotic vacuity consists in the undifferentiated nature of the modern, economic sign system: in contrast to genes, language, and even traditional, so-called multi-centric economies, modern money represents a code *with only one sign*. It signifies everything and nothing, lacking even a conventional relation (as in Peirce's definition of symbol) to any specific referent. A sign system with these properties is fundamentally at odds with the principles of organization in living systems, i.e. with life itself. As elaborated in Chapter 2, this reification and autonomization of exchange value has been well accounted for in the Marxian discussion of fetishism (cf. Taussig 1980), which is essentially a semiotic approach to political economy (Hornborg 2001b).

I have suggested that an ecosemiotic approach should lead us to ask questions about the roles of different kinds of human sign systems in transforming ecosystems. Semiotics, like any other useful methodology (e.g. logic, mathematics, cost-benefit analysis), always runs the risk of turning inward on itself so as to become its own, tautological obsession. When applied to the comprehension of empirical, cultural and ecological processes, however, it may help us reorganize our thinking in ways that permit new perspectives on old problems. The ongoing deterioration of the biosphere (often anthropocentrically referred to as our life-support systems or even natural capital) can thus be viewed as a problem of communication, deserving semiotic analysis. A semiotic approach appears to be the only conceivable framework for simultaneously considering the very diverse kinds of data on human–environmental relations that have been presented in this chapter, ranging from the phenomenology of unconscious sensations through structural analysis of elaborate linguistic constructions, to political economy. All these kinds of data are amenable to analysis founded on a general understanding of signs.

From this cursory discussion of the human ecology of Amazonia we may draw some general conclusions about the kinds of ecological transformations that can be attributed to different kinds of human sign systems. We have seen that human–environmental relations that are primarily mediated by direct, sensory and oral communication, e.g. subsistence horticulture and foraging, have tended to enhance biological diversity. We have reason to believe that ethnobiological vocabularies and metaphorical representations of human–environmental relations that are founded on intimate, practical experience of local ecosystems represent modes of calibration eminently suitable for negotiating the long-term co-evolution of human and non-human populations. An important ingredient in these modes of calibration is the inclination towards genuine communication with non-human interlocutors, which presupposes a dialogic stance founded on the animistic conviction that non-human Others are indeed sentient subjects. The effects of such communicative processes on ecosystems may indeed be transformative in the long run, but there is no evidence that they have ever threatened the integrity of Amazonian ecosystems as wholes.

Economic sign systems, on the other hand, have rapidly and drastically transformed human–environmental relations in Amazonia to the point where the entire rainforest ecosystem is under threat. Flows of reified exchange values had fundamentally transformed large areas of Amazonia already during the two millennia preceding European conquest. The devastating demographic impact of the arrival of Europeans to South America initially alleviated human designs on Amazonian ecosystems, so much so that the whole region was perceived to be virtually unaffected by human occupation. However, with the accelerating encroachment of colonialism, resource extraction, and the global market economy, ecological transformations were resumed at an unprecedented rate. In detaching themselves from the direct, face-to-face communication between humans and their natural environments, the fetishized flows of money and commodities – and the textualized knowledge systems that they engender – have no means of remaining geared to the long-term negotiation of local, ecological coexistence. On the contrary, their effect seems generally, in Amazonia as elsewhere, to have been to transform human perceptions of the environment from a community of sentient subjects to a mechanical assemblage of objects, thus truncating an ancient and literally vital dialogue. This objectivist stance has also been globally codified in the form of mainstream biological and ecological science. An ecosemiotic perspective would rather align itself with the animistic cosmologies of indigenous Amazonia, for perfectly rational reasons. It seems doubtful, however, that such local sensitivity to context can be significantly revitalized on a global scale without major transformations of the very idea of money.

8 Possible moneys and impossible machines

To intervene in the logic of capitalism

I will begin this final chapter by discussing some of the epistemological and existential circumstances that constrain our various approaches to sustainability. Further on, I will discuss what prospects we have of actually intervening in the destructive logic of capitalism addressed in the earlier chapters.

Over the years, I have on various occasions found myself drawn into three quite different kinds of discourses on the global ecological predicament of humanity. Within minutes of entering a meeting or lecture room, it would become obvious whether I was in the company of what I now call activists, managers, or analysts. These categories represent the fundamental points of departure for three separate approaches to sustainability.

1 For the activists, the foundational assumption is that global sustainability is a question of power, contradiction, and an engaging struggle against the adverse camp. Concepts such as environmental justice and political ecology well capture this stance, and my impression is that most of the students I encounter belong to this category.
2 For the analysts, global sustainability is a topic that should be illuminated through detached theoretical reflection and objective empirical research, as conducted, for instance, in the respectable field of Environmental History. The expectation is that it should be discussed from all kinds of perspectives, and in all kinds of detailed contexts, in a growing number of prestigious publications. Most committee members in research councils, deciding on the allocation of research grants, tend to belong to this category. (As teachers *and* researchers, of course, many of us thus find ourselves caught between two contradictory imperatives.)
3 Management, finally, seems to be the catchword or umbrella under which both these seemingly irreconcilable incentives can converge. The crucial expectation here is that any knowledge that is produced

> about sustainability should be *applicable*. Problems are interesting to discuss only if they can be solved. This discourse, towards which more and more students and research councils tend to gravitate, combines the students' ambition to change the world with the academic ambition to produce knowledge. But in gravitating toward management, as expressed in the catchword sustainable development, students tend to lose their interest in power and researchers their interest in reflexivity, because the implicit assumptions of management include consensus rather than contradiction, and applicability rather than truth. Thus, in subtle ways, the indignation of the students and the intellectual integrity of researchers are deflected and co-opted into acceptable problem-solving activities delineated by the way problems are politically defined. The language of policy and management tends to avoid questions of power, conflicts, and inequalities. Although conspicuously present – and increasingly problematic – in global human society, such issues are rarely identified as problems to be addressed by scientists.

These, then, are some of the constraining discursive spaces available in relation to issues of global sustainability, crisis, and alternatives in the early twenty-first century. Though intellectually lamentable, the politics of knowledge pushing us toward a discourse on management and policy is understandable, and academics have every reason to try to envisage ways of converting their insights into strategies for increasing sustainability.

The growing market for books with the word 'collapse' in their titles (e.g. Heinberg 2004; Diamond 2005) seems to suggest a widespread anticipation of imminent socio-ecological crisis. A very real issue for sustainability research is whether to rely on future crises to force core populations of the world system to transform their economic behaviour in a more sustainable direction, or whether the current preoccupation with collapse, and some of the more sustainable practices that it engenders, is itself a symptom of a crisis and transformation that is already under way.

As I have shown in earlier chapters, much of the current confusion regarding the prospects of sustainability derives from the miscommunication between social and natural science. In order to bring these two scholarly traditions together, it is essential to understand the history of their separation. Why have ecologists perceived ecosystems as devoid of human beings? Why have economists perceived societies as dissociated from their biophysical environment? Strategies for increasing sustainability need to be based on truly transdisciplinary knowledge, in the sense of acknowledging both the importance of fundamentally reorganizing societal flows of matter and energy *and* the importance of redesigning cultural sign systems such as language and money in order to achieve such reorganization.

Rather than addressing the history of Cartesian dualism, it may be more fruitful here to repeat our observations on the origins of the modern science of economics. What was the social context in which emerged the abstract vocabulary of economics? The contributions of one of its founding fathers, David Ricardo, illustrate how the discourse of economics was geared to accounting for the financial success of bankers and stockbrokers in the hub of the British colonial empire during the early decades of industrialism. Rather than attribute profits and economic growth in London to the tangible, biophysical inequalities of the so-called triangular trade between Britain, West Africa, and North America – which at times entailed grossly unequal exchanges of embodied labour and land – economic discourse was couched in abstractions such as utility, comparative advantage, and willingness to pay.

The study of the logic of money and market mechanisms became an end in itself, divorced from the material substance of production and consumption. This split had been observed already by Aristotle, who referred to the two perspectives as chrematistics versus *oikonomia* (cf. Martinez-Alier 1987). With modern, neoclassical economics, the obsession with chrematistics and monetary exchange value eclipsed all other considerations. No wonder that this paradigm is currently turning out to be an awkward tool for handling problems of global resource management. It was not designed for such a purpose, but to provide a way of managing monetary profit, irrespective of whether that profit had been derived from the Atlantic slave trade, Alabama cotton plantations built on slave labour and soil mining, or the Lancashire textile industry.

An economic science seriously concerned with the distant environmental, medical, and societal consequences of economic growth in a particular locale would need a completely different theoretical framework and vocabulary than conventional economics. In fact, to the extent that it has repeatedly proven unable to predict drastic changes even in financial flows – which should pose a serious dilemma for a science claiming to understand such matters – economics might have much to gain from taking the social and biophysical environment into consideration. Several critics of conventional economics have proposed that economic crises result precisely from the discrepancies between abstract wealth and material resources that economic science has chosen not to see (cf. Foster and Magdoff 2009; Harvey 2010).

If economic science in this sense qualifies as an ideology – a corpus of ideas that systematically mystifies social relations of power as if they were natural – it is all the more interesting to scrutinize precisely those aspects that economic analyses tend to exclude from view. This raises a long list of complex questions: How do local gains in efficiency in the use of time and space rely on appropriation and losses of time and space (embodied labour and land) elsewhere in the global system? How does the reified notion of market prices conceal asymmetric flows of embodied labour and land as if they were reciprocal exchanges? How can the technological capacity of a given population be represented as independent of that population's position in a global system of resource flows? How might we rethink notions of economic growth and technological development as a zero-sum game of global resource distribution, rather than a cornucopia? How do discrepancies between the economic and moral reach of individual market actors constrain their capacity to assume responsibility for the social and environmental consequences of their consumption? How are we as individuals persuaded to accept an ever more disturbing moral responsibility for the state of the world, while the power that reproduces this global order is increasingly centralized and inaccessible to us? How can we continue to accept only those analyses of global environmental problems that provide a comfortable solution? How can we learn to live with the ambiguity of accepting the historical probability of socio-ecological collapse without demanding simple and immediate solutions? How can crisis itself serve as an incentive to creatively generate more sustainable forms of economy, society, and culture? How, in redesigning cultural sign systems for sustainability, can we address the dimension of power inherent in current systems by reducing opportunities for accumulation, appropriation, and unequal exchange?

These are questions that I have tried to answer in several other contexts over the years. In the current context, however, I will be less preoccupied with analysing the epistemological and metabolic problems of industrial capitalism than with discussing the prospects for their 'solution'. What are the possibilities of radical changes in the currently unsustainable global organization of human–environmental relations? I will begin by briefly and critically discussing the solutions offered by mainstream proposals for sustainable development, in order to clear the ground for a more penetrating analysis. It is my conviction that the possibilities of sustainability are as theoretically demanding as the problems they aspire to solve.

Mainstream solutions and their problems

Let us initially recall that a significant proportion of both professional and amateur analysts of the global situation would deny that there are any problems that will not be solved by the continued expansion of the market and technology. As mentioned in Chapter 7, a prominent example of this stance is the Danish statistician Björn Lomborg (2001). Nevertheless, it is probably safe to assume that, a decade into the twenty-first century, most people who have seriously reflected about the course of fossil-fuelled capitalism will acknowledge the real dilemmas posed by climate change, peak oil, and the management of a limited global land area expected to feed a growing population, fuel its expanding technology, and preserve biodiversity. Once these dilemmas are acknowledged, it seems, visions of how to remedy the situation tend to follow one of two main options.

First, there is a very widespread conviction that solutions are to be found in new technologies. This position is pervasive throughout most of the literature and debate on sustainable development over the past decades (e.g. World Commission on Environment and Development 1987; Gore 1993; Brown 2008; Friedman 2008). To put this vision in a nutshell, it is based on a faith in the prospect of producing more goods with less environmental consequences. This strategy includes not only advocating tangible new technologies such as renewable energy and more efficient forms of architecture and land use, but more fundamentally a belief in practical engineering as a way to reorganize human behaviour, whether aiming to stabilize population, transform consumption patterns, reduce economic inequalities, improve health, or restore ecosystems. All these goals, of course, are eminently justifiable. It is as unthinkable to argue with them as to question the goal of sustainable development itself. But the question we need to put to ourselves, after three decades of costly rhetoric on sustainability, is whether devising schemes for achieving sustainability through engineering is really relevant, as long as it is becoming increasingly clear that the socio-ecological processes of industrial capitalism seem completely impervious to them? Are solutions really 'within reach' when economic logic (and political expediency) continues to reject them as impossible?

There are very good reasons to believe, on the contrary, that our current unsustainability is precisely what the inherent logic of industrial capitalism tends to promote. All the time, while spending billions of dollars on *talking* about sustainability, we have become less and less sustainable. Almost every single parameter conceivable – global per capita greenhouse gas emissions, energy use, resource consumption, ecological footprints, solid waste production, etc. – yields the same unequivocal conclusion. When after such a long time the solution continues to remain so far away from our reach, we must conclude that there is something lacking in the analysis. The faith in technology does not reckon with the inherent logic of our current socio-ecological system. To stand a chance of transforming that logic, we need to include incisive insights into its operation in our analysis of the problem. An example of major failure to do so is the current obsession of the European Union with imagining a technology for 'carbon capture and storage' that will encourage a vast expansion of coal combustion (Hansson 2008). The remarkably widespread belief in 'clean coal' is thus another example of the many impossible machines that inhabit our imagination about the future of industrial civilization. No more than nuclear reactors fuelled by water, or solar panels superior to photosynthesis, are such technological utopias likely to provide ecologically sound *and* economically feasible solutions for the world at large. As illustrated in earlier chapters, to advocate technological solutions to problems of sustainability is generally tantamount to suggesting environmental load displacement from wealthier to poorer segments of global society, since the new technologies are mystifications of unequal exchange and will only be available to the affluent.

A second main recipe for solving problems of sustainability is the belief in new ethics. If industrial capitalism cannot be expected to engage in sustainable production through its own autonomous logic, the idea is that it might be persuaded to do so if a sufficient number of people dedicate themselves to sustainable consumption. When there is a sufficient market for sustainably produced products, the argument goes, industry will adapt. This solution is often presented in combination with the preceding one, and can be criticized very much in the same way. The vision of responsible green consumers forcing industry to become sustainable has been proposed continuously over several decades, but has yielded very little in terms of changing global patterns of consumption and production. It seems to rest on the belief that moral conviction, education, and argument will ultimately yield six billion responsible and caring global citizens, willing to sacrifice more of their money, time, and comfort in solidarity with anonymous masses of people and ecosystems on other continents and in future centuries. Even if a minority of dedicated citizens in the affluent North were thus to create a niche for corporations offering genuinely sustainable products – and what is genuinely sustainable raises a number of additional problems – it is hardly to be expected that the low-wage masses of the global periphery should have the economic margins to permit their conscience to commit them to green consumption. In acknowledging the huge discrepancy between the socio-ecological and ethical reach of the average modern citizen, it would thus be more realistic to propose a reduction of the former than an expansion of the latter.

Alternatives to collapse: intervening in the logic of capitalism?

Among those for whom neither of these two strategies – technology or ethics – appears plausible, there is the option to seriously anticipate socio-environmental *collapse*. There are plenty of precedents in world history

(Tainter 1988). Increasingly, the prospect of collapse – whether financial, technological, ecological, social, political, or a combination of these – is being discussed not only as a threat, but as a possibility. It is quite possible that nothing short of a system breakdown will force us to change our habits. Although it may seem inappropriate to think of collapse as a kind of solution, this argument has gained considerable momentum in the context of current concerns about peak oil and global warming (e.g. Heinberg 2004). Generally speaking, to view contradiction and crisis as a possibility for renewal is a recurrent approach in the study of both societal and psychological systems (cf. Schneider *et al.* 2010). In the social sciences it would be both irresponsible and uninformed to deny the possibility that global industrial capitalism may be facing socio-ecological collapse.

Of the alternatives mentioned so far – technology, ethics, and collapse – none is primarily based on what humans can collectively decide through political decisions. What might a politics for sustainable development look like, and what could it achieve? This is a complex consideration with several levels, since political decisions could be used to enforce almost any strategy for sustainability, depending on the scale of political decision-making (accountable to local, national, or global electorates), its immediate context of events, discourses, and incentives, and on what is at stake for individual politicians. So far, there is little to indicate that political decisions will decisively influence the sustainability of either production or consumption at any level, but it is quite conceivable that the political options will look very different in a situation where socio-ecological collapse seems an imminent reality. Until then, we may observe that the threat of collapse has generally been used to legitimize political decisions that are most questionable from the perspective of world democracy (Chomsky *et al.* 2002; Klein 2007). This has become increasingly clear as global competition over natural resources has hardened in recent years (Klare 2001; Harvey 2003; Bunker and Ciccantell 2005). In other words, crises generated by the logic of industrial capitalism are represented so as to enhance our faith in this very capitalism. Against this background, we have every reason to be wary of the strategies now being devised to evade the consequences of climate change, resource shortages, and other widely acknowledged constraints on the current organization of global metabolism. Although socio-ecological collapse would certainly seem to vindicate much of the green criticism of industrial capitalism, there will no doubt be more than one way of interpreting such a crisis, yielding very different political recipes.

Given an open discussion about the real nature of global socio-ecological crisis, and an electorate that is not bribed to believe that short-term national interests in the long run can prevail over global imperatives, it is not completely inconceivable that political decision-making could in fact promote more sustainable ways of organizing the world economy and metabolism. But in order to begin to envisage what such changes would entail, we first need to consider what industrial capitalism *is* and some examples of attempts to confront its inherent logic.

This is not the place to engage in extended theoretical discourse about the historical specificity of industrial capitalism. Suffice to say that its specificity may reside in a recursive expansion of money and machines. There are several perspectives on this specificity, ranging from some world-system theorists who argue that basically similar forms of capital accumulation can be traced five millennia back in time (Frank and Gills 1993) to more or less rigid exegesis of Karl Marx's *Das Kapital* (1867–1893). Neither of these approaches seems very useful as we consider the current socio-ecological dilemma of global capitalism. There are certainly continuities between pre-industrial and industrial modes of accumulation (Hornborg 2000), but no one can deny that the Industrial Revolution in the eighteenth century, or even the advent of capitalist agriculture in the sixteenth (Wallerstein 1974), represents an important *dis*continuity. In any case, it is reasonable not to expect Karl Marx in the 1860s to have predicted the global socio-metabolic logic of capital accumulation as it has unfolded 150 years later. Since then, important new perspectives have emerged on the roles of land, labour, and capital in global metabolism (see, for instance, Jorgenson and Clark 2009). Suffice it to say that, in industrial capitalism, monetary profits can be invested in technology, which in turn can yield additional profits. This is a very simple baseline on which I believe most of us can agree. There is, in other words, a recursive (or positive feedback) relation between economic and technological growth, i.e. between money and machines.

Another way of expressing this would be to say that, in the modern world, money and machines are two concrete manifestations of capital, or even of capital*ism*. Money is converted into machines (i.e. technology), and machines yield more money. Let us for the moment leave aside the issue of how such processes of technological accumulation are ecologically unsustainable or entail unequal net transfers of quantifiable resources such as energy, materials, embodied land, or embodied labour (Jorgenson and Clark 2009), in order to focus instead on how their concrete manifestations – as money objects and actual machines – are *perceived* by the human beings whose lives and social relations are configured by these objectifications. Money and machines are both fetishes in the sense that they are objectified social relations: relations between human beings masquerading as relations between things. As previously observed, both assume the appearance of objects endowed with autonomous, generative powers, and in the process mystifying social relations of unequal exchange. In order to fully understand our predicament and to stand a chance of evading collapse, we shall have to expose the mystified social relations of exchange that present themselves to us as money, commodities, and machines.

In Chapter 2, we discussed the Luddite movement, which attacked the new machinery that was being accumulated in the industrial districts of England 200 years ago. We saw that to confront capitalism by resorting to physical sabotage against machinery and its owners has historically proved a dead end. Physical sabotage does not appear to be a viable strategy for confronting capital. Nor does the revolutionary assault on the owners of such capital, as long as the underlying mechanisms of capital accumulation continue to resurface in regime after regime. If rejecting machines is not a feasible strategy for counteracting the logic of capitalism, however, the final word has not yet been said about the prospect of rejecting *money*.

The root of all evil: confronting money from Mammon to alternative currency movements

The sociological, philosophical, and even theological analysis of what the idea and institution of money does to human life, social relations, and human–environmental relations is a vast field. The literature is enormous and impossible to review, but the preoccupation with this most potent of human inventions stretches from the biblical denunciation of Mammon to present-day concerns with alternative currency movements. Few would deny the pivotal role of money in shaping modern societies and their environmental impacts, yet few would seriously question whether it should continue to have this role, even among the majority of people who acknowledge the risk of global socio-ecological collapse. In part this may simply be because they have not thought about money as an institution about which we even theoretically have a choice, in part because the idea of choosing to abandon it seems utterly naïve.

But perhaps transforming our money system is the only chance we have. General-purpose money rewards the dissipation of resources with ever more resources to dissipate, until they are gone, or at least inaccessible. The dilemma of sustainability thus seems to be the very juxtaposition of this socio-cultural institution with the Malthusian facts of entropy, limited land area, and finite stocks of resources. The problem could thus be expressed as the consequences of money in a universe obeying the Second Law of Thermodynamics. If this is indeed recognized as our fundamental problem, it is much less problematic to conclude which of these factors – general-purpose money or the Second Law of Thermodynamics – can be changed through political decisions. Money is a cultural sign system invented by humans and in the long run perhaps the only factor we can hope to transform in the interest of sustainability. In the next section, we shall consider what options are available to us, viewed through the lens of economic anthropology.

Why a sustainable economy would have to be multi-centric

Fifty-five years ago, Paul Bohannan (1955) published a ten-page article in the *American Anthropologist* that was destined to become a classic in economic anthropology. It was titled *Some Principles of Exchange and Investment among the Tiv*, and built on twenty-six months of fieldwork in northern Nigeria between July 1949 and January 1953. The Tiv people with whom Bohannan worked were at that time mainly subsistence farmers who, in Bohannan's words, prided 'themselves on their farming abilities and their subsistence wealth'.

The traditional Tiv economy, however, is remembered by anthropology students primarily because of what Bohannan referred to as its multi-centric character. Bohannan observed that the Tiv recognized three separate *spheres of exchange* that were hierarchically arranged, so that transactions implying conversions between them were considered morally charged.[1] To convert food into prestige goods, or prestige goods into women, was considered skilful – a measure of success – but to convert in the opposite direction, that is, women into prestige goods, or prestige goods into food, was morally reprehensible. Bohannan's analysis showed two things very clearly: first, that it is quite conceivable to organize the economy according to a moral hierarchy of values, where all values are *not* commensurable, and second, that such morally differentiated economies will quickly collapse when incorporated into the modern realm of general-purpose money.

The introduction of modern currency, writes Bohannan, 'was hurried by the Administration' through the imposition of a money tax by the end of the First World War. One of the most tangible consequences of the introduction of general-purpose money among the Tiv was, in fact, the sudden increase in food exports, leaving Tiv elders in the 1950s cursing money itself for 'depriving them of their food faster than they are able to increase production'. Obviously, then, the moral semiotics of money – in other words, how exchange is culturally conceived – can have very significant material implications. After twenty-six months of fieldwork,

Bohannan was acutely aware of the detrimental impacts of general-purpose money on subsistence security among the Tiv, but concludes the article by observing that 'the ethnographer can only look on and attempt to understand the ideas and motivations', knowing that the traditional system of ideas will inevitably be 'smashed' in the confrontation with modern money.

The expansion of a global market economy, as experienced by the Tiv and by humanity as a whole, has been fuelled by the tenets of neoliberal economics. As we have seen, these tenets of economics were conceived by hugely successful bankers and stockbrokers such as David Ricardo, situated over 150 years ago in the hub of the British colonial empire, constructing models and concepts with which to account for their own success in accumulating capital. Their models and concepts were obviously *not* designed to alleviate global inequalities, or problems of sustainability, yet they continue to frame and constrain modern economic discourse on these issues – to the extent that they are at all addressed. Anthropologists have the choice of subscribing to this hegemonic discourse, or to approach it precisely as Bohannan approached the economic categories of the Tiv – as a cultural system. There have been several attempts to do this over the years, notably by Stephen Gudeman (1986), but a lot of work remains, particularly if we want to say something that sounds relevant to the ongoing deliberations on fair and sustainable trade.

Over the years, I have found Stephen Gudeman's work very useful as a means of relativizing (or defamiliarizing; cf. Marcus and Fischer 1986) standard categories of economics. In particular, his book *Economics as Culture: Models and Metaphors of Livelihood* (1986) showed how different categories of people in various historical and geographical contexts – ethnographically documented Africans and Melanesians as well as nineteenth-century British economists and their modern descendants – have conceptualized exchange and material provisioning in radically different but ubiquitously *cultural* ways. To any science founded – like economics – on the authority of its privileged perspective, this should be a powerfully subversive message. Gudeman did make a distinction between local/metaphorical models and universal/derivational/mathematical (i.e. Western) models, but his point was finally to transcend this distinction by showing that universal models are 'another set of local formulations' (1986:viii), the special rationale of which is the aspiration to subsume and eclipse other local models (ibid., 154–157). Imperialism, he concluded, may be rethought of in terms of 'who gets to model whom'. This conclusion to me suggests a final, reflexive victory of substantivism over formalism[2] in economic anthropology, applying cultural relativism so as to expose and challenge the political agenda of master narratives in economics. It represents the post-modern turn in anthropology at its best, while it still had something serious and politically subversive to say.

Neoliberal economic theory is a reflection of the logic of general-purpose money. General-purpose money is in itself basically an *idea* about the generalized interchangeability of all things. In making products and services from all over the world commensurable in terms of a single metric, the nineteenth-century world economy vastly increased the opportunities for – and the scope of – unequal exchange. In Chapter 5, I showed that, in 1850, by exchanging, on the world market, £1,000 worth of cotton textiles for £1,000 worth of raw cotton from its colonies, Britain in fact traded 4,092 hours of British labour for 32,619 hours of overseas labour, and the use of less than a hectare of British land for the use of 58.6 hectares of land overseas. These measures refer to the labour and land embodied in the commodities exchanged, i.e. the real material resources that went into their production. The ideologically most significant aspect of the neoliberal economic world-view that Ricardo helped to consolidate around this time was precisely to make such real material resources invisible. From now on, there was only utility, quantifiable in money. That is why mainstream economists to this day are uncomfortable with the concept of unequal exchange. Market transactions are by definition equal, if monetary prices are the only metric with which they are gauged. I would argue that, although submerged, this is one of the very points of Ricardo's constructions.

In growing increasingly wealthy on the stock exchange, Ricardo could hardly have been expected to congratulate himself on exploiting suffering slaves and deteriorating soils in Alabama or India. At this particular historical moment, looking out at the world through a window in Britain, a man of finance was compelled to dispel all hesitations about unequal exchange; to count money value as the one and only measure of substance; to view land, labour, and capital as mutually substitutable; and to hail free trade as the very foundation of economic growth. Ricardo's science of economics is a cultural construction, as Gudeman (1986) observed twenty-five years ago, but as such it is also profoundly ideological. In order to pave the way for industrial capitalism, it had to make unequal exchange of labour and land invisible, by conditioning us to think only in terms of money. It was compelled to create the illusion that machines (capital) can substitute for land and labour, rather than acknowledge that capital draws on the appropriation of land and labour from elsewhere in the world system. And it needed to make us all believe that the invisible hand of free trade operates in the best interests of everybody, and that any attempt to constrain the economy by appeal to moral principles – as had the Tiv previously – should be dismissed as backwardness and repression.

The idea of general-purpose money is a successful vehicle of capital accumulation for exactly the same reason that it is, as the Tiv and so many others have observed, the root of all evil. In today's world, it is this idea of generalized interchangeability that generates the polarization between economic and technological growth, on one hand, and economic and ecological impoverishment, on the other. The simple logic of general-purpose money tends to accelerate processes of resource extraction and consumption by rewarding the transformation of resources into commodities with ever more resources to transform. The more cotton textiles that Britain could export in 1850, for example, the more raw cotton it could import – which means that the more slaves could be put to work, the more soils could be degraded, and so on. The same logic underlies our economic successes and global ecological vulnerabilities today: the more industrial products that the USA, Europe, and Japan can export today, the more resources they will be able to dissipate tomorrow. Or, in even simpler words: the more oil we dissipate today, the more oil we will be able to dissipate tomorrow. This is the simple algorithm of destruction that results from the operation of general-purpose money in a universe obeying the Second Law of Thermodynamics (Hornborg 2001b). These are tangible, material problems for those of us concerned, for example, with global inequalities in energy use or the unevenly distributed consequences of global warming, but, although as intellectually incomprehensible as it is ideologically predictable, they do not seem to pose a problem for neoliberal economics.

When lecturing about these matters and arriving at this rather pessimistic conclusion on the future of industrial capitalism, there is always someone who asks me to say something about what could possibly be done to prevent global economic and ecological collapse. In those situations I often find myself referring to Paul Bohannan and the multi-centric economy of the Tiv. If our fundamental problem is the very conjunction of the idea of general-purpose money and the Second Law of Thermodynamics, we need to take seriously the fact that only one of these conditions can be transformed through political decisions. We cannot change the Entropy Law, but it is entirely possible to transform the idea and institution of money. I am not suggesting, of course, that we should try to reinstate the specific spheres of exchange recognized fifty years ago by the Tiv, but that we could draw inspiration from the very notion of recognizing a moral hierarchy of incommensurable values. If the politically enforced introduction of general-purpose money in northern Nigeria a mere sixty years ago was able to so completely reorganize the conditions for sustainable subsistence, it is obvious that similarly profound changes – in another direction – could be triggered by other kinds of politically enforced transformations of the money system. I am thus convinced that, sooner or later, there will be a renaissance for the multi-centric economy.

Anthropologists tend to maintain a healthy distance to social engineering, but as stewards of a vast collective memory of global cultural experimentation, it could be argued that they have an obligation to share these alternative ideas and ideologies with non-anthropologists seeking visions of a sustainable world. Like so many other populations over the past few centuries, the Tiv have discovered how their incorporation into a world economy orchestrated by general-purpose money implied loss of local control, inequitable trade relations, and environmental degradation. It is no longer as obvious as it seemed sixty years ago that such problems are only the necessary and temporary side effects of an inevitably linear path toward modernization and development (Ferguson 1999). It may be high time to vindicate those Tiv elders who clearly perceived how the dismantling of their economic categories served to pave the way for external appropriation of their resources.

In order to immunize, as it were, local land and labour against the ravages of global capital, it may well prove necessary to insulate them culturally, as resource categories that are not reducible to the universal metric of general-purpose money. Rather than reversing history, such a scenario might be perceived as a way of reflexively taking stock of humankind's historical experience of several centuries of globalization. Minimally, this would mean distinguishing – in cultural, moral, legal, and practical terms – between those essential goods and services that are basic to survival, and that can generally be locally produced and distributed, on one hand, and those essential contributions of modernity (such as computer software and advanced pharmaceuticals) that *require* a global division of labour, on the other. Only by reinstating some kind of moral hierarchy distinguishing such separate spheres of exchange can the algorithm of destruction be domesticated. The building blocks of world society – the communities and ecosystems that we so often evoke by referring to the 'local' – need to be granted their own integrity, and a measure of autonomy, in order to begin to be able to engage in global trade on terms that could be called fair. Hundreds of anthropological case studies from all over the world have shown that the logic of general-purpose money is *not* conducive to such local integrity.

Economic anthropology is, thus, a repository of empirically and theoretically well-founded arguments for a critical reconsideration – and reshaping – of the institution of money itself. I need not recapitulate the history of ideas from Marx, Mauss, Polanyi, Dalton, Bohannan, Godelier, Sahlins, Rappaport, Taussig, Bloch, Kopytoff, and Gudeman to make this point. Suffice to say that many anthropologists, in comparing non-modern modes of economic life with those of capitalist modernity, have discovered both the destructiveness and the cultural arbitrariness of the notion of general-purpose money. In various ways, they have emphasized the inclination of most people, irrespective of cultural background, to recognize that all values are not commensurable and interchangeable. This moral conviction is obvious even where no attempts are made to challenge the idea of general-purpose money or in other ways reshape the institutions that govern our economic life, for instance in standard reactions to stories of illegal trade in human organs. At the same time, people all over the world have been expressing such convictions through their material practice of exchange, whether affluent middle-class students in the North pursuing a sense of community by joining barter clubs, or impoverished families in the South resorting to the 'informal' economy in order to survive. Although their incentives are extremely diverse, their economic strategies converge in distinguishing the local and personal sphere of exchange as more secure and morally committing than the global and abstract.

There is a growing literature on community currencies, barter clubs, and other experiments in creating local spheres of exchange (e.g. Dobson 1993; Douthwaite 1999; North 2007), but there have been very few attempts to understand such social projects in terms of anthropological theory. Gudeman (2001:160–161) briefly mentions these experiments while arguing for a distinction between the principles of market and community, but does not seriously consider their potential as a way to alleviate the socio-political polarizations and ecological problems generated by the capitalist world market. A radical anthropological discussion of the prospects for a transformation of the idea of money is Keith Hart's (2001) *Money in an Unequal World.* Hart, whose experiences of 'development' in Ghana in the 1960s prompted him to coin the expression 'informal economy' (Hart 1992), suggests that the Internet could be used to make it possible for each individual to use different *kinds* of money for transactions within different social networks. The new capacity for electronic information storage would make it possible to 'personalize' money and credit so as to record and express the specific transaction history of each actor. Hart does not pursue this idea to the point of offering a concrete, practical blueprint for such a personalized economy in which we all 'make our own money', but he suggests that local currencies such as LETS (Local Exchange Trading Systems) and Ithaca dollars are examples of what the future will bring (Hart 2001:279–285).

Although what Hart proposes is explicitly subsumed under Bohannan's category of special-purpose money, he curiously chooses to dismiss the logic of spheres of exchange as the preoccupation of Tiv elders and other 'aristocracy' with their rank privileges (ibid., 271–272). This is perplexing, as he is later (ibid., 301) very clear about the democratic advantages of such special currencies:

> They can give communities or networks of individuals a means of organizing some of their activities independently, without fear that the value of their transactions will somehow be sucked off to an anonymous center of accumulation.

Hart's book is refreshingly outspoken in its criticism of what he calls state capitalism, but its recipes for alleviating the problems are contradictory. The argument is entrenched in the same kind of technological optimism that prevented Marx from facing the implications of his own conclusion that the machines in themselves are manifestations of unequal exchange (Hornborg 2001b). Hart makes a point of distinguishing markets from capitalism, criticizing the latter as inherently exploitative but suggesting that markets are a problem only when some people have inordinately more money than others. He does not, however, specify how his scheme of multiple currencies might curb such uneven accumulation of money through, for instance, investments in technology. Although he observes that uneven mechanization is fundamental to accumulation, he suggests that to conflate the Marxian concept of money capital with machines is to mystify the social relations involved (Hart 2001:84). I would counter that the real mystification is to *not* recognize technologies as manifestations of social relations of exchange.

An important merit with Hart's book, however, is that it shows very convincingly how dominant modern powers, such as the United States and the European Union, behind all the neoliberal discourse on free trade, practice a very 'unfree' trade policy, particularly as regards agricultural produce. The recognition or conviction that all things are not freely interchangeable, it seems, is as fundamental to nation-states as to individual market actors. This fact could potentially have surprising and far-reaching implications for current deliberations on free trade. Rather than continue to let their agricultural subsidies contradict their neoliberal discourse, an interesting option for these nation-states might be to openly concede that the land (or landscape) within their territories can legitimately be regarded as the base of a special sphere of exchange that is exempt – and needs to be protected – from free, global competition. Land and its products is not a kind of capitalist commodity like any other. It has not been difficult for the United States and the European Union to come up with good arguments for preserving their rural areas – socially and ecologically – rather than let them succumb to competition from economically more efficient producers, whether the latter's 'efficiency' derives from lower labour costs, more lenient environmental legislation, or other advantages. It is thus ironic that these powers should practice, at home, an anti-market arrangement that they have consistently denied peoples of the South such as the Tiv.

If, in the light of these considerations, we should wish to follow Hart in crossing the boundary between anthropological theory and the engineering of radical social policy, we could begin by asking ourselves, how might a multi-centric modern economy be devised? A major justification for such a reform would be to replace current agricultural policies with an economic framework that makes it rational for individual humans to practice sustainable production and consumption. This goal would, in turn, entail a conscious encouragement of increasingly localized flows of energy, materials, and nutrients, i.e. reducing long-distance transports and making extensive recycling more attractive and feasible than at present.

Let us think about the various functions of money (Table 8.1). Stephen Gudeman (2008:124–136) distinguishes between money as means, as medium, and as end. The former, exemplified by so-called scrip, is a means for settling debts and used in one-directional transactions such as exchanging coupons or welfare stamps for goods. Scrip is special-purpose money in the sense that it specifies the objects or services for which it can be exchanged. Medium or commercial money, on the other hand, is 'used in two-way exchanges, is not limited to particular things and services, and permits a wide range of choice in use' (ibid., 134). Finally, financial or interest-yielding money is an end in itself, since access to it will give the owner incentives to accumulate capital. Modern money combines the latter two functions, serving both as a general medium of exchange and as a store of self-aggrandizing value, or capital.

In Table 8.1, I have analysed the possible functions of money in a slightly different way than Gudeman. In thinking about how money could be transformed into an institution for promoting sustainability – rather than resource depletion and socio-ecological collapse – we could imagine a kind of money (let us call it 'Notes') that combines the first and the second

Table 8.1 The functions of money (functions of Notes shared with other currencies marked in bold)

Functions of money	*$€£*	*LETS*	*Scrip*	*Notes*
Medium of two-directional transactions	**Yes**	**Yes**	No	**Yes**
Store of interest-yielding value	Yes	**No**	**No**	**No**
Specifies goods and services to be exchanged	No	No	**Yes**	**Yes**
Specifies range of users	**No**	Yes	**No**	**No**
Issued and guaranteed by state authorities	**Yes**	No	No	**Yes**

of these functions rather than the second and the third. In distinguishing between money as means and money as medium, Gudeman does not consider the logical alternative of a money system that combines the special-purpose character of scrip and the two-way character of commercial money. Theoretically, it would be quite possible to design such a currency, which can be used for certain kinds of exchanges but does not yield interest and encourage accumulation. For some reason, however, none of the many historical and recent experiments with alternative currencies that I am aware of has exploited this particular possibility (North 2007). All of them – e.g. German stamp scrip, Coventry Social Credits, Manchester bobbins, Ithaca dollars, Budapest Talentum, New Zealand's Green Dollars, Argentinian *créditos*, Brixton pounds – seem to have been founded on the idea of restricting the circulation of money in space, but without systematically restricting its circulation in kind, i.e. its interchangeability or applicability. But precisely this, I believe, is the very Gordian knot of sustainability.

The so-called LETS experiments (Local Exchange Trading Systems) are based on the idea of encouraging local and egalitarian exchanges of goods and services without the mediation of formal, financial money. However, they do not delimit the sphere of goods and services to be exchanged by this means. It is thus theoretically possible to exchange a loaf of home-baked bread from Brixton for computer software from California. This amorphousness of LETS deprives them of much of the emancipatory potential of the idea of alternative currencies. The generally limited success and longevity of these movements may in fact partly derive from an insufficiency of their guiding visions and an absence of rigorous analysis.

A fundamental point of departure for alternative currency movements ought to be the recognition that all values are not commensurable. If the rationale of LETS networks is to strengthen local community, the definition of 'local' should be taken more seriously. It is reasonable to expect local values to pertain to goods and services that can indeed be provided locally, and even more fundamentally, to goods and services that are essential for local community survival. For most of humanity, these two kinds of local values tend to largely coincide. Food, shelter, and basic care can usually be provided fairly locally, as they have been for millennia, whereas most of the luxuries and capital goods that absorb the major share of our modern salaries are long-distance imports. Let us, for the sake of simplicity, call these two categories of commodities local goods and global goods, respectively. I am convinced that, in the long run, sustainability will hinge on our capacity to acknowledge that these two categories are incommensurable spheres of exchange. If it were possible to insulate, so to speak, the local reproduction of community from the global flows of capital, currently prevalent patterns of energy use and other resource management would be profoundly transformed. Not only would human societies become more sustainable in an ecological sense, but the resurrection of community would also give the notion of *social* sustainability a very tangible significance. Vulnerability to risk of several kinds would be reduced. If we are serious about the global necessity of frugality, recycling, and reducing our ecological footprints, it is incumbent on us to think very seriously about the kinds of social arrangements that might encourage such changes. If the fundamental problem of sustainability is the generalized interchangeability of all values on the world market, which rewards an accelerating dissipation of resources with access to more resources to dissipate, a logical solution would be to restrict the interchangeability of values by introducing separate currencies for separate spheres of exchange.

Changing the logic of money: a vision

Let us conduct a thought experiment. To reduce transports, oil dependency, and carbon dioxide emissions, a nation-state wishes to encourage its citizens to increasingly consume local produce. It thus issues a special currency (notes) that can only be used in exchange for a clearly specified range of products and services that we may call local in the dual sense that they are locally produced and essential for basic subsistence and survival. These local products and services would include, for instance, basic foodstuffs, organic fuels, and building materials deriving from the local landscape, as well as fundamental community services such as care of children and elderly, simpler health care, and assistance with carpentry, local transports, and practical chores of various kinds. The Notes issued by the state for use within this local or informal sphere are distributed to households in proportion to their size, and the households are free to use them according to their own priorities. No transactions with Notes are taxed, nor can accumulation of Notes yield any interest.

What would motivate households to actually use their Notes, rather than throw them away? The answer is that they would liberate a substantial part of their formal income, formerly used to purchase local (or potentially local) goods and services, for other purposes. Throwing them away with coupons and discount offers would literally be like throwing away money. Initially, the introduction of a local currency might thus in fact increase consumption of non-local goods. Households would be able to choose to which extent they wish to continue to earn a part of their income from the formal sector, but most households could be expected to devote part of their labour time to the formal sector and part to the informal, or local.

What would motivate shopkeepers, farmers, foresters, nurses, carpenters, etc. to accept Notes as payment for their products and services, rather than demand payment in formal currency? The income in Notes is tax-free and can be employed for a number of purposes, in addition to what has been mentioned, including flexibly hiring informal labour (without costly, longer-term commitments) and purchasing goods and services of importance for their own enterprise, e.g. household compost for farmers and used packaging materials for shopkeepers. Considering rising costs of energy and transports, exchange rates in Notes for local produce could be expected to provide households and local enterprise with more beneficial transaction opportunities than offered by corresponding commodities on the formal market. In addition, the state could offer to exchange a specified proportion of earned Notes for formal money at beneficial exchange rates, thereby closing the circulation of Notes. The idea, in other words, would be to create an alternative market for tax-free, two-way exchanges of local goods and services actively encouraged by the authorities.

Why would the state want to subsidize and encourage an increase in tax-free transactions at the local level? The answer is that the indirect benefits in environmental, social, and medical terms, and in terms of reduced costs e.g. for transport infrastructure, would exceed the reduction of tax revenue. Let me give a few examples. If there was a general shift toward more localized consumption, the reduction of transports, energy use, and greenhouse gas emissions would have very significant economic and environmental effects. If nutrients and packaging materials were largely locally recycled, it would reduce waste, resource use, and the eutrophication of water. If agriculture became more geared to local demand, and had access to untaxed local labour, it would shift to less mechanized production, further reducing resource use and environmental damage. If the demand for long-distance imports of agricultural products decreased, it would alleviate the marginalization of local populations from commercially exploited land worldwide, increasing their chances of gaining a local livelihood and experiencing food security. If food production became more localized, it would reduce waste of food through discarded surplus production and losses in storage and transport, while encouraging the production of fresher and healthier food, with fewer preservatives. If there was a resurgence of local agriculture, it would encourage a more complex pattern of global land use, increasing biodiversity in areas now used for industrial monocultures as well as in areas marginalized from agricultural production. If local self-sufficiency in products and services essential to survival was thus increased, it would increase global food security and reduce communities' vulnerability to crisis. If local community was thus revitalized through an intensified exchange of local goods and services, it would also reduce social marginalization through what we now know as 'unemployment' and improve health among citizens otherwise prone to such marginalization.

In all these ways, the introduction of an alternative currency as defined here would reduce the currently escalating administrative and technological costs of the logistic, social, medical, and ecological problems associated with our present pattern of organizing the global economy. The costs of a constantly expanding infrastructure for solving all these problems, not to mention infrastructure for long-distance transports, evoke what Tainter (1988) calls 'declining returns on complexity', a recurrent dilemma of societies caught in vicious circles leading to socio-ecological collapse. From a more optimistic perspective, and at a more profound, existential level, the conscious and organized dismantling of much of the contemporary infrastructure for transports and commerce would improve the capacity of individual consumers to grasp the extent of their societal and ecological agency, and to assume responsibility for it. It would encourage them to develop identities based more on authentic social relations than on the semiotics of consumption.

Against the background of serious fiscal crises shaking several developed nations in the first decade of the twenty-first century, the idea of a state-led shift toward increasing localization of societal metabolism through the introduction of alternative currencies can be viewed as aligned with current calls for what has become known as *degrowth* (cf. Schneider *et al.* 2010; Jackson 2009; Victor 2008). The perceived realism of this proposal can probably be expected to increase with the severity of such financial crises in the years ahead.

And what would happen to capitalist enterprise and the world market? No doubt a great many people would be forced to adapt to completely new circumstances, abandoning some life projects while pursuing others. But is not this precisely what the capitalist world market has been coercing us to do, continuously, for centuries, from the Luddites of early nineteenth-century Yorkshire to the unemployed in modern Detroit? The difference would be that this particular transition would be for the good of all, that is, for a truly sustainable development, and that the resultant bi-centric economy would immunize us in the future from many of the risks that have previously accompanied transitions. The formal market can be expected to adapt, as it has always done. It would thrive in those supra-local spheres of exchange where it will continue to be useful and in demand, for instance in sectors such as information technology and pharmaceuticals. The world market is definitely not the most efficient way of managing global energy resources, food production, or farmland, but if delimited to the specific function of providing access to vital goods and services that cannot be locally produced, it may still prove crucial for the integration of world society. By seeing to it that all significant social actors would benefit from a carefully redesigned and politically sanctioned money system, finally making it economically rational to be sustainable, the problems of political acceptance currently obstructing any efficacious sustainability policy can be overcome. There are undoubtedly plenty of potential problems with this admittedly utopian scheme. My simple response is that these problems can hardly be more difficult to handle than the ones we are presently facing.

To make more tangible how commodity fetishism has severed the connections between ecology, economy, and social morality (community), let me conclude with a fictional anecdote from a rural setting anywhere in the northern hemisphere: A local, unemployed woman in her fifties struggles to improve her modest income by selling blueberries to the affluent, middle-class professionals who have settled in her community, within comfortable commuting distance to their urban offices. She has braved the heat and rain and mosquitoes for days on end to deliver those ten litres of blueberries for the fifteen dollars they will fetch. The buyers deposit them in their freezer, but when one of their teenagers shifts the plug for another electrical appliance, she forgets to reconnect the freezer and the contents are ruined. When reprimanded by her parents, she retorts that there is no reason to quarrel over 'mere material things'.

The contemporary disarticulation of consumption from the resource flows and inputs of human labour that sustain them keeps economy and morality – things and persons – effectively apart. Such commodity fetishism rests on the invisibility of production. Its implications can sometimes protrude clearly in rural economies where this disarticulation is not always guaranteed. A consciously reorganized local sphere of exchange might thus increase our ability to grasp the social and ecological consequences of our consumption, and to reintroduce morality into economic life.

Notes

Introduction

1 By 'semiotic processes' I mean the ways in which different people attribute meanings to the things and events that surround them.

1 Zero-sum world: how to think about ecologically unequal exchange

1 The notion of 'underpaid' labour or land may be misleading, however, since it suggests that the basic problem is insufficient monetary compensation. If the terms of trade were to approach reciprocity in such material terms, there would simply be no incentive to continue the exchange. Unequal terms of trade, viewed from a biophysical perspective, are thus a prerequisite for much of the exchange and production that we think of as 'modern' and 'developed'.

2 Fetishism, dissociation, and the cultural analysis of capitalism

1 For a more systematic discussion of the various uses of the concept of fetishism, see Hornborg 2001a.
2 I use the word 'machines' to refer to technological objects the existence and operation of which is ultimately dependent on access to inanimate energy sources such as fossil fuels and electricity.
3 The invalidity of the Cartesian dichotomy challenged by 'actor network theory' (ANT), by imputing agency to material objects, has been recognized as applying generally to human interaction with the remainder of the biosphere (Latour 2004; Bennett 2010). There is definitely a sense in which material objects of various kinds – even landscapes – can be agents, prompting or constraining human action. However, agency does not necessarily mean subjectivity or intentionality, while the agency of technological artefacts that we here refer to as 'machines' generally deserves to be scrutinized in terms of the embodiment or delegation of political intentions.
4 Probably because animism would imply such moral constraints, the few Western scientists who have seriously championed an animistic world-view (e.g. Bateson 1972; Uexküll 1982 [1940]) have inexorably been relegated to the margins. This is not because their arguments about the semiotic and communicative dimension

of ecosystems have been shown to be invalid, but because they have been found irrelevant to the modern project. The primary interest of Western science is not to get to know living organisms as subjects, but as objects.

5 This disjunction of technology and economy is equally characteristic of mainstream studies in the philosophy of technology (for an overview, see Ihde 1993).

6 Cf. Pietz's (1985–1988) remark that colonial European writers frequently observed that pre-modern Africans were inclined to perceive technological objects as magical beings.

7 Compare Latour's (2010) discussion of iconoclasm.

8 In fact, it is not unusual to regard Marx as, in some respects, a victim of the technological ideology of nineteenth-century industrialism (e.g. Benton 1989:76).

9 The etymology of the word 'fetish' can be traced through a Portuguese word for magic and sorcery (*feitico*) to a Latin word for manufacture (*facticius*; Pietz 1985–1988).

10 This has in fact been suggested by Ruyle (1977:223–225), but we should recall that cannibalism here serves as a metaphor rather than a literal understanding of consumption. Ruyle does not distinguish between the consumption of food and the consumption of other 'material use-values', only the former of which is actually consumed in the sense of being incorporated in the physical body of the consumer, but only the latter of which are produced in the sense of incorporating energy expended by the producer.

3 Historical political ecology: progress as environmental load displacement

1 As noted in Chapter 1, however, this suggestion has been extensively criticized by John Bellamy Foster and Paul Burkett (Foster and Burkett 2004; Burkett and Foster 2006, 2008).

4 Towards a truly global environmental history

1 Compare, however, McNeill 2010 for such a pertinent perspective on the 'success' of the contemporary Netherlands.

2 See Chapter 6 for a more elaborate discussion of the relation between concepts of 'value' and unequal exchange.

5 The unequal exchange of time and space

1 The genealogy of the notion of 'ecological footprints' goes back to Borgström's (1965) concept of 'ghost acreages' (cf. also Catton 1980). Pomeranz's (2000:313–315) calculation of the ghost acreages of some British imports (sugar, timber, cotton) in the late eighteenth and early nineteenth century was brought to my attention after this text had been written, but rests on the same general interpretation of British history (cf. Wilkinson 1973).

2 I have added the number of working-hours required for production of one ton of woollen manufactures (6,968) and the number of hours required for production of 1.33 ton of raw wool (1,314), in order to reckon with 25 per cent loss of weight in processing.

3 I have estimated the number of hectares required for production of 1.33 ton of raw wool (36.8), in order to reckon with 25 per cent loss of weight in processing.
4 The reason why no attempt is made here to include capital in these calculations is that capital is itself understood as embodied labour and land, i.e. the product of past relations of unequal exchange of time and space. As such, any attempt to justify the rates of exchange discussed here by reference to British capital would merely serve to mystify the unequal process of accumulation that the calculations are intended to reveal.
5 20,346 – 4,092.
6 It should also be noted that agricultural production of fibres such as cotton, flax, and hemp are more demanding on the soil, and on labour, than most other crops (Pomeranz 2000:218–219), which made this environmental load displacement particularly advantageous from the British perspective.
7 To generate an additional £66,475,547 from sales of woollen manufactures in 1850 would have required an additional output of 194,372 tons (× £342), representing 217,696 tons of raw wool demanding 6,030,196 hectares of land.
8 We may note that the very same Cartesian conception of material nature that is often implicated in deliberations on the significance of European cosmology for 'technological progress' now serves to impede the struggle of social sciences to grasp the phenomenon of technology itself.
9 In fact, even steam engines 'of a sort' had been developed by non-European societies, including China, before the eighteenth century (Pomeranz 2000:61–62, citing Needham 1965).
10 The combustion of fossil fuels not only means relying on 'ghost acreages' of the distant past, it reduces future access to such fuels and generates future problems with carbon dioxide emissions, ocean acidification, and climate change (cf. Roberts and Parks 2007).
11 This is not always evident even to non-Eurocentric historians. Thus, Pomeranz (2000:57) refers to 'the breakthroughs in chemistry that today allow capital to substitute for land (and labor) to an astonishing degree (above all through using synthetic fertilizer…)', but without mentioning that synthetic fertilizers require fossil fuels and finite stocks of phosphates. Industrial technologies for improving soil fertility thus tend to reproduce the 'ecological imperialism' of the guano trade (cf. Clark and Foster 2009).
12 The total arable land of Britain at this time was roughly seventeen million acres (Pomeranz 2000:275).
13 Pomeranz (2000:276, n.50) believes that twenty-one million 'ghost acres' would be a more correct figure. For further calculations of the 'ghost acreages' represented by fossil fuels (including oil), see Catton 1980.
14 Building on figures provided by Pomeranz (2000:274–278, 313–315), we can conclude that the 'ghost acreages' available to Britain in the year 1900 through imports of sugar and cotton alone amounted to over 195 million acres. In the same year, the combustion of coal would have equalled 210 million acres of woodland. These three commodities together in 1900 thus represented 405 million acres of ecologically productive land. The total land surface of Great Britain is about fifty-nine million acres.

6 Value, unequal exchange, and uneven development

1 Kenneth Hermele, personal communication, 2011.

2 Kenneth Hermele, personal communication, 2010.

7 Vital signs: how money transforms ecosystems

1 Such a dialectic between images of society and images of nature can also be identified in the history of Western civilization, as is obvious in the ideological affinities between capitalism and Darwinism, and in the dialectic between kin-based or political authority, on the one hand, and images of a patriarchal divinity, on the other. Other examples might include the reciprocal mirroring of a clock-like, rationally constructed society, on the one hand, and a mechanistic view of the cosmos, on the other.

8 Possible moneys and impossible machines: to intervene in the logic of capitalism

1 Whether the Tiv economy was accurately described by Bohannan, which has been debated in anthropology, is not at issue here as much as the very idea of a multi-centric economy.
2 For an accessible introduction to the formalist–substantivist controversy, see Wilk and Cliggett 2007.

References

Adams, R. McC. 1996. *Paths of fire: An anthropologist's inquiry into Western technology*. Princeton: Princeton University Press.

Adas, M. 1989. *Machines as the measure of men: Science, technology, and ideologies of Western dominance*. Ithaca: Cornell University Press.

Adas, M. 2006. *Dominance by design: Technological imperatives and America's civilizing mission*. Cambridge, Mass.: The Belknap Press of Harvard University Press.

Andersson, J. O. 2010. International trade in a full and unequal world. In *International trade and environmental justice: Toward a global political ecology*. A. Hornborg and A. K. Jorgenson, eds. New York: Nova Science Publishers, pp. 113–125.

Andersson, J. O. and M. Lindroth 2001. Ecological unsustainable trade. *Ecological Economics* 37:113–122.

Apter, E. and W. Pietz, eds. 1993. *Fetishism as cultural discourse*. Ithaca: Cornell University Press.

Århem, K. 1991. Ecosofía Makuna. In *La selva humanizada: Ecología alternativa en el trópico húmedo colombiano*. F. Correa, ed. Bogotá: ICAN/FEN/CEREC, pp. 105–122.

Århem, K. 1996. The cosmic food web: Human-nature relatedness in the Northwest Amazon. In *Nature and society: Anthropological perspectives*. P. Descola and G. Pálsson, eds. London: Routledge, pp. 185–204.

Arvelo-Jiménez, N. and H. Biord 1994. The impact of conquest on contemporary indigenous peoples of the Guiana Shield: The System of Orinoco Regional Interdependence. In *Amazonian Indians from prehistory to the present: Anthropological perspectives*. A. C. Roosevelt, ed. Tucson: University of Arizona Press, pp. 55–78.

Bairoch, P. 1993. *Economics and world history: Myths and paradoxes*. Chicago: University of Chicago Press.

Balée, W. 1989. Managed forest succession in Amazonia: The Ka'apor case. In *Resource management in Amazonia: Indigenous and folk strategies*. Advances in Economic Botany 7. D. A. Posey and W. Balée, eds. Bronx: New York Botanical Garden, pp. 129–158.

Balée, W. 1993. Indigenous transformation of Amazonian forests: An example from Maranhão, Brazil. L'Homme 126/128:231–254.
Balée, W. 1995. Historical ecology of Amazonia. In *Indigenous peoples and the future of Amazonia: An ecological anthropology of an endangered world*. L. E. Sponsel, ed. Tucson: University of Arizona Press, pp. 97–110.
Balée, W. and C. L. Erickson, eds. 2006. *Time and complexity in historical ecology: Studies in the Neotropical lowlands*. New York: Columbia University Press.
Bates, D. G. and S. H. Lees, eds. 1996. *Case studies in human ecology*. New York: Plenum.
Bateson, G. 1972. *Steps to an ecology of mind*. Frogmore: Paladin.
Bateson, G. 1979. *Mind and nature: A necessary unity*. New York: Dutton.
Baudrillard, J. 1972. *Pour une critique de l'economie politique du signe*. Paris: Gallimard.
Baudrillard, J. 1975. *The mirror of production*. St. Louis: Telos.
Bennett, J. 2010. *Vibrant matter: A political ecology of things*. Durham: Duke University Press.
Benton, T. 1989. Marxism and natural limits: An ecological critique and reconstruction. *New Left Review* 178:51–86.
Berkes, F. 1999. *Sacred ecology: Traditional ecological knowledge and resource management*. Philadelphia: Taylor & Francis.
Berkes, F. and C. Folke, eds. 1998. *Linking social and ecological systems: Management practices and social mechanisms for building resilience*. Cambridge: Cambridge University Press.
Berkes, F., J. Colding, and C. Folke, eds. 2003. *Navigating social-ecological systems: Building resilience for complexity and change*. Cambridge: Cambridge University Press.
Berlin, B. 1992. *Ethnobiological classification: Principles of categorization of plants and animals in traditional societies*. Princeton: Princeton University Press.
Bettelheim, C. 1972. Theoretical comments. In *Unequal exchange: A study of the imperialism of trade*. A. Emmanuel. New York: Monthly Review Press, pp. 271–322.
Binfield, K., ed. 2004. *Writings of the Luddites*. Baltimore: The Johns Hopkins University Press.
Bird-David, N. 1993. Tribal metaphorization of human-nature relatedness. In *Environmentalism: The view from anthropology*. K. Milton, ed. London: Routledge, pp. 112–125.
Bird-David, N. 1999. 'Animism' revisited: Personhood, environment, and relational epistemology. *Current Anthropology* 40(1):S67–91.
Blaug, M. 1961. The productivity of capital in the Lancashire cotton industry during the nineteenth century. *Economic History Review* 13:358–381.
Blaut, J. M. 1993. *The colonizer's model of the world: Geographical diffusionism and Eurocentric history*. New York: The Guilford Press.
Blaut, J. M. 2000. *Eight Eurocentric historians*. New York: The Guilford Press.
Bloch, M. 1989. The symbolism of money in Imerina. In *Money and the morality*

of exchange. J. Parry and M. Bloch, eds. Cambridge: Cambridge University Press, pp. 165–190.

Bohannan, P. 1955. Some principles of exchange and investment among the Tiv. *American Anthropologist* 57:60–70.

Borgström, G. 1965. *The hungry planet: The modern world at the edge of famine*. New York: Macmillan.

Boserup, E. 1988. Environment, population, and technology in primitive societies. In *The ends of the Earth: Perspectives on modern environmental history*. D. Worster, ed. Cambridge: Cambridge University Press, pp. 23–38.

Bourdieu, P. 1984. *Distinction: A social critique of the judgement of taste*. London: Routledge.

Braudel, F. 1992 [1979]. *Civilization and capitalism, 15th–18th centuries*, 3 vols. Berkeley: University of California Press.

Brolin, J. 2006. *The bias of the world: Theories of unequal exchange in history*. PhD dissertation. Lund Studies in Human Ecology 9. Lund University.

Brondízio, E. S. 2006. Landscapes of the past, footprints of the future: Historical ecology and the study of contemporary land-use change in the Amazon. In *Time and complexity in historical ecology: Studies in the Neotropical lowlands*. W. Balée and C. L. Erickson, eds. New York: Columbia University Press, pp. 365–405.

Brown, L. R. 2008. *PLAN B 3.0: Mobilizing to save civilization*. Washington: Earth Policy Institute.

Bryant, R. L. and S. Bailey 1997. *Third world political ecology*. London: Routledge.

Bunker, S. G. 1985. *Underdeveloping the Amazon: Extraction, unequal exchange, and the failure of the modern state*. Chicago: University of Chicago Press.

Bunker, S. G. 2007. Natural values and the physical inevitability of uneven development under capitalism. In *Rethinking environmental history: World-system history and global environmental change*. A. Hornborg, J. R. McNeill and J. Martinez-Alier, eds. Lanham: AltaMira Press, pp. 239–258.

Bunker, S. G. and P. S. Ciccantell 2005. *Globalization and the race for resources*. Baltimore: The Johns Hopkins University Press.

Burkett, P. and J. B. Foster 2006. Metabolism, energy, and entropy in Marx's critique of political economy: Beyond the Podolinsky myth. *Theory and Society* 35:109–156.

Burkett, P. and J. B. Foster 2008. The Podolinsky myth: An obituary introduction to 'Human labour and unity of force', by Sergei Podolinsky. *Historical Materialism* 16:115–161.

Camino, A. 1977. Trueque, correrías e intercambios entre los Quechuas andinos y los Piro y Machiguenga de la montaña peruana. *Amazonía Peruana* 1(2):123–140.

Carneiro, R. L. 1974 [1970]. Hunting and hunting magic among the Amahuaca of the Peruvian montaña. In *Native South Americans: Ethnology of the least known continent*. P. Lyon, ed. Boston: Little, Brown, pp. 122–132.

Catton, W. R., Jr. 1980. *Overshoot: The ecological basis of revolutionary change*. Urbana: University of Illinois Press.

Chagnon, N. A. 1968. *Yanomamö: The fierce people*. New York: Holt, Rinehart & Winston.

Chew, S. C. 2001. *World ecological degradation: Accumulation, urbanization, and deforestation 3000 BC–AD 2000*. Lanham: AltaMira Press.

Chew, S. C. 2006. *The recurring Dark Ages: Ecological stress, climate changes, and system transformation*. Lanham: AltaMira Press.

Chomsky, N., P. R. Mitchell, and J. Schoeffel 2002. *Understanding power: The indispensable Chomsky*. New York: The New Press.

Clark, B. and J. B. Foster 2009. Ecological imperialism and the global metabolic rift: Unequal exchange and the guano/nitrates trade. *International Journal of Comparative Sociology* 50(3–4):311–334.

Clark, B. and R. York 2005. Carbon metabolism: Global capitalism, climate change, and the biospheric rift. *Theory and Society* 34:391–428.

Clark, G. 1993. Agriculture and the Industrial Revolution, 1700–1850. In *The British Industrial Revolution*. J. Mokyr, ed. Boulder: Westview Press, pp. 227–266.

Clark, R. P. 1997. *The global imperative: An interpretive history of the spread of humankind*. Boulder: Westview Press.

Clastres, P. 1977. *Society against the state*. Oxford: Basil Blackwell.

Collins, E. J. T. 1969. Harvest technology and labour supply in Britain, 1790–1870. *Economic History Review* 22:453–473.

Costanza, R. 1980. Embodied energy and economic evaluation. *Science* 210:1219–1224.

Croll, E. and D. Parkin, eds. 1992. *Bush base-forest farm: Culture, environment and development*. London: Routledge.

Cronon, W. 1983. *Changes in the land: Indians, colonists, and the ecology of New England*. New York: Hill and Wang.

Cronon, W. 1991. *Nature's metropolis: Chicago and the Great West*. New York: W.W. Norton.

Crosby, A. W. 1972. *The Columbian exchange: Biological and cultural consequences of 1492*. Santa Barbara: Greenwood Press.

Crosby, A. W. 1986. *Ecological imperialism: The biological expansion of Europe, 900–1900*. Cambridge: Cambridge University Press.

Crumley, C. L., ed. 1994. *Historical ecology: Cultural knowledge and changing landscapes*. Santa Fe: School of American Research Press.

Culbert, T. P. 1973. The downfall at Tikal. In *The Classic Maya collapse*. T. P. Culbert, ed. Albuquerque: University of New Mexico Press, pp. 63–92.

Culbert, T. P. 1988. The collapse of Classic Maya civilization. In *The collapse of ancient states and civilizations*. N. Yoffee and G. L. Cowgill, eds. Tucson: University of Arizona Press, pp. 69–101.

Deane, P. 1957. The output of the British woolen industry in the eighteenth century. *The Journal of Economic History* 17:207–223.

Debeir, J.-C., J.-P. Deléage, and D. Hémery 1991 [1986]. *In the servitude of power: Energy and civilization through the ages*. London: Zed Books.

Descola, P. 1992. Societies of nature and the nature of society. In *Conceptualizing Society*. A. Kuper, ed. London: Routledge, pp. 107–126.

Descola, P. 1994. *In the society of nature: A native ecology in Amazonia*. Cambridge: Cambridge University Press.

Descola, P. and G. Pálsson, eds. 1996. *Nature and society: Anthropological perspectives*. London: Routledge.

Diamond, J. 1997. *Guns, germs, and steel: The fates of human societies*. New York: W.W. Norton.

Diamond, J. 2005. *Collapse: How societies choose to fail or succeed*. New York: Viking.

Dobson, R. V. G. 1993. *Bringing the economy home from the market*. Montreal: Black Rose Books.

Douthwaite, R. 1999. *The ecology of money*. Foxhole: Green Books/The Schumacher Society.

Eisenmenger, N. and S. Giljum 2007. Evidence from societal metabolism studies for ecological unequal trade. In *The world system and the Earth system: Global socioenvironmental change and sustainability since the Neolithic*. A. Hornborg and C. Crumley, eds. Walnut Creek: Left Coast Press, pp. 288–302.

Ellen, R. F. 1982. *Environment, subsistence and system: The ecology of small-scale social formations*. Cambridge: Cambridge University Press.

Ellen, R. F. 1988. Fetishism. *Man* (N.S.) 23:213–235.

Ellen, R. F. 1996. The cognitive geometry of nature: A contextual approach. In *Nature and society: Anthropological perspectives*. P. Descola and G. Pálsson, eds. London: Routledge, pp. 103–123.

Ellen, R. and K. Fukui, eds. 1996. *Redefining nature: Ecology, culture and domestication*. Oxford: Berg.

Emmanuel, A. 1972. *Unequal exchange: A study of the imperialism of trade*. New York: Monthly Review Press.

Erickson, C. L. 2006. The domesticated landscapes of the Bolivian Amazon. In *Time and complexity in historical ecology: Studies in the Neotropical lowlands*. W. Balée and C. L. Erickson, eds. New York: Columbia University Press, pp. 235–278.

Erickson, C. L. and W. Balée 2006. The historical ecology of a complex landscape in Bolivia. In *Time and complexity in historical ecology: Studies in the Neotropical lowlands*. W. Balée and C. L. Erickson, eds. New York: Columbia University Press, pp. 187–233.

Eriksen, L. 2011. *Nature and culture in prehistoric Amazonia: Using G.I.S. to reconstruct ancient ethnogenetic processes from archaeology, linguistics, geography, and ethnohistory*. PhD dissertation. Lund Studies in Human Ecology 12. Lund University.

Evernden, N. 1985. *The natural alien*. Toronto: University of Toronto Press.

Fabian, J. 1983. *Time and the Other: How anthropology makes its object*. New York: Columbia University Press.

Farnie, D. A. 1979. *The English cotton industry and the world market 1815–1896*. Broadbridge: Clarendon Press.

Ferguson, J. 1999. *Expectations of modernity: Myths and meanings of urban life on the Zambian copperbelt*. Berkeley: University of California Press.

Fischer-Kowalski, M. 1998. Society's metabolism: The intellectual history of material flow analysis, part I: 1860–1970. *Journal of Industrial Ecology* 2(1):61–78.

Fischer-Kowalski, M. and C. Amann 2001. Beyond IPAT and Kuznets curves: Globalization as a vital factor in analysing the environmental impact of socio-economic metabolism. *Population and Environment* 23(1):7–47.

Fischer-Kowalski, M. and H. Haberl 1993. Metabolism and colonization: Modes of production and the physical exchange between societies and nature. *Innovation: The European Journal of Social Sciences* 6(4):415–442.

Fisher, W. H. 2007. Surrogate money, technology, and the expansion of savanna soybeans in Brazil. In *Rethinking environmental history: World-system history and global environmental change*. A. Hornborg, J. R. McNeill, and J. Martinez-Alier, eds. Lanham: AltaMira Press, pp. 345–360.

Foster, J. B. 1999. Marx's theory of metabolic rift. *American Journal of Sociology* 105(2):366–405.

Foster, J. B. 2000. *Marx's ecology: Materialism and nature*. New York: Monthly Review Press.

Foster, J. B. and P. Burkett 2004. Ecological economics and classical Marxism. *Organization & Environment* 17(1):32–60.

Foster, J. B. and P. Burkett 2008. Classical Marxism and the second law of thermodynamics: Marx/Engels, the heat death of the universe hypothesis, and the origins of ecological economics. *Organization & Environment* 21(1):3–37.

Foster, J. B. and Clark, B. 2004. Ecological imperialism: The curse of capitalism. In *Socialist register 2004: The new imperial challenge*. L. Panitch and C. Leys, eds. London: Merlin Press, pp. 186–201.

Foster, J. B. and F. Magdoff 2009. *The great financial crisis: Causes and consequences*. New York: Monthly Review Press.

Foster, J. B., B. Clark and R. York 2010. *The ecological rift: Capitalism's war on the Earth*. New York: Monthly Review Press.

Fowler, C. S. 1977. Ethnoecology. In *Ecological anthropology*. D. L. Hardesty, ed. New York: John Wiley & Sons, pp. 215–243.

Fox, N. 2002. *Against the machine: The hidden Luddite tradition in literature, art, and individual lives*. Washington: Island Press.

Frank, A. G. 1966. The development of underdevelopment. *Monthly Review* 18:17–31.

Frank, A. G. 1967. *Capitalism and underdevelopment in Latin America*. New York: Monthly Review Press.

Frank, A.G. 1978. *World accumulation 1492–1789*. New York: Monthly Review Press.

Frank, A. G. 1995. The modern world system revisited: Rereading Braudel and Wallerstein. In *Civilizations and world systems: Studying world-historical change*. S. K. Sanderson, ed. Lanham: AltaMira Press, pp. 163–194.

Frank, A. G. 1998. *ReOrient: Global economy in the Asian age*. Berkeley: University of California Press.

Frank, A. G. 2007. Entropy generation and displacement: The nineteenth-century multilateral network of world trade. In *The world system and the Earth system: Global socioenvironmental change and sustainability since the Neolithic*. A. Hornborg and C. Crumley, eds. Walnut Creek: Left Coast Press, pp. 303–316.

Frank, A. G. and B. K. Gills, eds. 1993. *The world system: Five hundred years or five thousand?* London: Routledge.

Friedel, R. 2007. *A culture of improvement: Technology and the Western millennium.* Cambridge, Mass.: MIT Press.

Friedman, J. 1974. The place of fetishism and the problem of materialist interpretations. *Critique of Anthropology* 1:26–62.

Friedman, J. 1979. Hegelian ecology: Between Rousseau and the world spirit. In *Social and ecological systems.* P. C. Burnham and R. F. Ellen, eds. London: Academic Press, pp. 253–270.

Friedman, T. L. 2008. *Hot, flat, and crowded: Why we need a green revolution – and how it can renew America.* New York: Farrar, Straus and Giroux.

Gade, D. W. 1972. Comercio y colonización en la zona de contacto entre la sierra y las tierras bajas del valle del Urubamba en el Perú. *XXXIX Congreso Internacional de Americanistas*, vol. 4. Lima, pp. 207–221.

Georgescu-Roegen, N. 1971. *The Entropy Law and the economic process.* Cambridge, Mass.: Harvard University Press.

Gierlinger, S. 2010. Die langfristigen Trends der Material- und Energieflüsse in den USA in den Jahren 1850 bis 2005. Institut für Soziale Ökologie, Universität Klagenfurt.

Gills, B. K. and W. R. Thompson, eds. 2006. *Globalization and global history.* London: Routledge.

Godelier, M. 1969. La monnaie de sel des Baruya de Nouvelle-Guinée. *L'Homme* 9(2):5–37.

Godelier, M. 1986. *The mental and the material.* New York: Verso.

Goldfrank, W. L., D. Goodman, and A. Szasz, eds. 1999. *Ecology and the world-system.* Santa Barbara: Greenwood Press.

Gore, A. 1993. *Earth in the balance: Ecology and the human spirit.* New York: Plume.

Graeber, D. 2001. *Toward an anthropological theory of value: The false coin of our own dreams.* Houndmills: Palgrave.

Graeber, D. 2007. *Possibilities: Essays on hierarchy, rebellion, and desire.* Oakland: AK Press.

Gudeman, S. 1986. *Economics as culture: Models and metaphors of livelihood.* London: Routledge & Kegan Paul.

Gudeman, S. 2001. *The anthropology of economy: Community, market and culture.* Oxford: Blackwell.

Gudeman, S. 2008. *Economy's tension: The dialectics of community and market.* Oxford: Berghahn.

Håkansson, N. T. 2004. The human ecology of world systems in East Africa: The impact of the ivory trade. *Human Ecology* 32(5):561–591.

Håkansson, N. T. 2007. Trade, 'trinkets', and environmental change at the edge of world-systems: Political ecology and the East African ivory trade. In *Rethinking environmental history: World-system history and global environmental change.* A. Hornborg, J. R. McNeill, and J. Martinez-Alier, eds. Lanham: AltaMira Press, pp. 143–162.

Hammer, O. 2000. *Claiming knowledge: Strategies of epistemology from theosophy to the New Age*. PhD dissertation, Department of Theology, Lund University.

Hanegraaff, W.J. 1998. *New Age religion and Western culture: Esotericism in the mirror of secular thought*. New York: State University of New York Press.

Hanley, N. 1998. Resilience in social and economic systems: A concept that fails the cost-benefit test? *Environment and Development Economics* 3:244–249.

Hansson, A. 2008. *The return of coal: Carbon dioxide capture and storage in science and politics*. PhD dissertation. Linköping Studies in Arts and Science No. 436. Linköping University.

Harris, M. 1979. *Cultural materialism: The struggle for a science of culture*. New York: Vintage.

Hart, K. 1992. Market and state after the Cold War: The informal economy reconsidered. In *Contesting markets: Analyses of ideology, discourse and practice*. R. Dilley, ed. Edinburgh: Edinburgh University Press, pp. 214–227.

Hart, K. 2001. *Money in an unequal world*. New York: Texere.

Harvey, D. 1996. *Justice, nature, and the geography of difference*. Oxford: Blackwell.

Harvey, D. 2003. *The new imperialism*. Oxford: Oxford University Press.

Harvey, D. 2010. *The enigma of capital and the crises of capitalism*. Oxford: Oxford University Press.

Headrick, D. R. 2010. *Power over peoples: Technology, environments, and Western imperialism, 1400 to the present*. Princeton: Princeton University Press.

Heelas, P. 2008. Spiritualities of life: New Age romanticism and consumptive capitalism. Oxford: Blackwell.

Heinberg, R. 2004. *Power down: Options and actions for a post-carbon world*. Gabriola Island: New Society Publishers.

Hettne, B. 1990. *Development theory and the three worlds*. Burnt Mill: Longman.

Heyman, J. 2007. Environmental issues at the U.S.–Mexico border and the unequal territorialization of value. In *Rethinking environmental history: World-system history and global environmental change*. A. Hornborg, J. R. McNeill, and J. Martinez-Alier, eds. AltaMira Press, pp. 327–343.

Hirsch, E. and M. O'Hanlon, eds. 1995. *The anthropology of landscape: Perspectives on place and space*. Oxford: Oxford University Press.

Hobsbawm, E. J. 1968. *Industry and empire: An economic history of Britain since 1750*. London: Penguin.

Hoffmeyer, J. 1996. *Signs of meaning in the universe*. Bloomington: Indiana University Press.

Holling, C. S. 1978. *Adaptive environmental assessment and management*. Hoboken: Wiley.

Hornborg, A. 1988. *Dualism and hierarchy in lowland South America: Trajectories of indigenous social organization*. PhD dissertation. Uppsala Studies in Cultural Anthropology 9. Stockholm: Almqvist & Wiksell.

Hornborg, A. 1989. Imperialismens termodynamik. *Tvärsnitt* 4:32–41.

Hornborg, A. 1992. Machine fetishism, value, and the image of unlimited good: Toward a thermodynamics of imperialism. *Man* (N.S.) 27:1–18.

Hornborg, A. 1993. Panoan marriage sections: A comparative perspective. *Ethnology* 32:101–108.

Hornborg, A. 1996. Ecology as semiotics: Outlines of a contextualist paradigm for human ecology. In *Nature and society: Anthropological perspectives*. P. Descola and G. Pálsson, eds. London: Routledge, pp. 45–62.

Hornborg, A. 1998a. Serial redundancy in Amazonian social structure: Is there a method for poststructuralist comparison? In *Transformations of Kinship*. M. Godelier, T. R. Trautmann, and F. E. Tjon Sie Fat, eds. Washington: Smithsonian Institution Press, pp. 168–186.

Hornborg, A. 1998b. Towards an ecological theory of unequal exchange: Articulating world system theory and ecological economics. *Ecological Economics* 25(1):127–136.

Hornborg, A. 1999a. Money and the semiotics of ecosystem dissolution. *Journal of Material Culture* 4(2):143–162.

Hornborg, A. 1999b. Comment on Nurit Bird-David, 'Animism' revisited: Personhood, environment, and relational epistemology. *Current Anthropology* 40(1):S80–81.

Hornborg, A. 2000. Accumulation based on symbolic versus intrinsic 'productivity': Conceptualizing unequal exchange from *Spondylus* shells to fossil fuels. In *World system history: The social science of long-term change*. R. Denemark, J. Friedman, B. Gills, and G. Modelski, eds. London: Routledge, pp. 235–252.

Hornborg, A. 2001a. Symbolic technologies: Machines and the Marxian notion of fetishism. *Anthropological Theory* 1(4):473–496.

Hornborg, A. 2001b. *The power of the machine: Global inequalities of economy, technology, and environment*. Walnut Creek: AltaMira Press.

Hornborg, A. 2005. Ethnogenesis, regional integration, and ecology in prehistoric Amazonia. *Current Anthropology* 46(4):589–620.

Hornborg, A. 2007. Conceptualizing socioecological systems. In *The world system and the Earth system: Global socioenvironmental change and sustainability since the Neolithic*. A. Hornborg and C. Crumley, eds. Walnut Creek: Left Coast Press, pp. 1–11.

Hornborg, A. and C. Crumley, eds. 2007. *The world system and the Earth system: Global socio-environmental change and sustainability since the Neolithic*. Walnut Creek: Left Coast Press.

Hornborg, A. and J. D. Hill, eds. 2011. *Ethnicity in ancient Amazonia: Reconstructing past identities from archaeology, linguistics, and ethnohistory*. Boulder: University Press of Colorado.

Hornborg, A., J. R. McNeill, and J. Martinez-Alier, eds. 2007. *Rethinking environmental history: World-system history and global environmental change*. Lanham: AltaMira Press.

Hughes, J. D. 1994. *Pan's travail: Environmental problems of the ancient Greeks and Romans*. Baltimore: Johns Hopkins University Press.

Hughes, J. D. 2001. *An environmental history of the world: Humankind's changing role in the community of life*. London: Routledge.

Ihde, D. 1993. *Philosophy of technology: An introduction*. New York: Paragon.

Ihde, D. and E. Selinger, eds. 2003. *Chasing technoscience: Matrix for materiality*. Bloomington: Indiana University Press.

Ingerson, A. E. 1994. Tracking and testing the Nature-Culture dichotomy. In *Historical ecology: Cultural knowledge and changing landscapes*. C. L. Crumley, ed. Santa Fe: School of American Research Press, pp. 43–66.

Ingold, T. 1986. *The appropriation of nature: Essays on human ecology and social relations*. Manchester: Manchester University Press.

Ingold, T. 1992. Culture and the perception of the environment. In *Bush base-forest farm: Culture, environment and development*. E. Croll and D. Parkin, eds. London: Routledge, pp. 39–56.

Ingold, T. 1996. Hunting and gathering as ways of perceiving the environment. In *Redefining nature: Ecology, culture and domestication*. R. Ellen and K. Fukui, eds. Oxford: Berg, pp. 117–155.

Ingold, T. 2000. *The perception of the environment: Essays in livelihood, dwelling and skill*. London: Routledge.

Jackson, J. B. C., M. X. Kirby, W. H. Berger *et al.* 2001. Historical overfishing and the recent collapse of coastal ecosystems. *Science* 293:629–638.

Jackson, T. 2009. *Prosperity without growth: Economics for a finite planet*. London: Earthscan.

Jones, C. 1979. Tikal as a trading center: Why it rose and fell. Paper presented at the XLIII International Congress of Americanists, Vancouver.

Jones, S. E. 2006. *Against technology: From the Luddites to neo-Luddism*. London: Routledge.

Jorgenson, A. K. 2003. Consumption and environmental degradation: A cross-national analysis of the ecological footprint. *Social Problems* 50(3):374–394.

Jorgenson, A. K. and B. Clark, eds. 2009. Special Issue: Ecologically unequal exchange in comparative perspective. *International Journal of Comparative Sociology* 50(3–4).

Jorgenson, A. K. and E. Kick, eds. 2006. *Globalization and the environment*. Leiden: Brill.

Jorgenson, A. K. and J. Rice 2005. Structural dynamics of international trade and material consumption: A cross-national study of the ecological footprints of less-developed countries. *Journal of World-Systems Research* 11(1):57–77.

Jorgenson, A. K. and J. Rice 2007. Uneven ecological exchange and consumption-based environmental impacts: A cross-national investigation. In *Rethinking environmental history: World-system history and global environmental change*. A. Hornborg, J. R. McNeill, and J. Martinez-Alier, eds. Lanham: AltaMira Press, pp. 273–288.

Kearney, M. 1984. *World view*. Novato: Chandler & Sharp.

Kelekna, P. 1981. Achuara food taboos. In *Food taboos in lowland South America*. K. M. Kensinger and W. H. Kracke, eds. Working Papers on South American Indians 3. Bennington: Bennington College, pp. 177–185.

Kensinger, K. M. and W. H. Kracke, eds. 1981. *Food taboos in lowland South America*. Working Papers on South American Indians 3. Bennington: Bennington College.

Kirch, P. V. and T. L. Hunt, eds. 1997. *Historical ecology in the Pacific Islands: Prehistoric environmental and landscape change*. New Haven: Yale University Press.

Klare, M. T. 2001. *Resource wars: The new landscape of global conflict*. New York: Owl Books/Henry Holt.

Klein, N. 2007. *The shock doctrine: The rise of disaster capitalism*. New York: Picador.

Kopytoff, I. 1986. The cultural biography of things: Commoditization as process. In *The social life of things: Commodities in cultural perspective*. A. Appadurai, ed. Cambridge: Cambridge University Press, pp. 64–91.

Kormondy, E. J. and D. E. Brown 1998. *Fundamentals of human ecology*. Upper Saddle River: Prentice Hall.

Kracke, W. H. 1981. Don't let the piranha bite your liver: A psychoanalytic approach to Kagwahiv (Tupí) food taboos. In *Food taboos in lowland South America*. Kensinger, K. M. and W. H. Kracke, eds. Working Papers on South American Indians 3. Bennington: Bennington College, pp. 91–142.

Kracke, W. H., ed. 1993. *Leadership in lowland South America*. South American Indian Studies 1. Bennington: Bennington College.

Krausmann, F., S. Gingrich, N. Eisenmenger *et al.* 2009. Growth in global materials use, GDP and population during the 20th century. *Ecological Economics* 68(10):2696–2705.

Kull, K. 1998. Semiotic ecology: Different natures in the semiosphere. *Sign Systems Studies* 26: 344–371.

Lathrap, D. W. 1968. The 'hunting' economies of the Tropical Forest Zone of South America: An attempt at historical perspective. In *Man the hunter*. R. B. Lee and I. DeVore, eds. Chicago: Aldine, pp. 23–29.

Lathrap, D. W. 1973. The antiquity and importance of long-distance trade relationships in the moist tropics of Pre-Columbian South America. *World Archaeology* 5:170–186.

Lathtrap, D. W. 1974. The moist tropics, the arid lands, and the appearance of great art styles in the New World. *Special Publications of the Museum of the Texas Technological University* 7:115–158.

Latour, B. 1993. *We have never been modern*. Cambridge, Mass.: Harvard University Press.

Latour, B. 1996. *Aramis or the love of technology*. Cambridge, Mass.: Harvard University Press.

Latour, B. 2004. *Politics of nature: How to bring the sciences into democracy*. Cambridge, Mass.: Harvard University Press.

Latour, B. 2010. *On the modern cult of the factish gods*. Durham: Duke University Press.

Leach, E. 1964. Anthropological aspects of language: Verbal categories and animal abuse. In *New directions in the study of language*. E. Lenneberg, ed. Cambridge, Mass.: MIT Press, pp. 25–63.

Lélé, S. 1998. Resilience, sustainability, and environmentalism. *Environment and Development Economics* 3:249–254.

Levin, S. A., S. Barrett, S. Aniyar *et al.* 1998. Resilience in natural and socioeconomic systems. *Environment and Development Economics* 3:222–235.

Lévi-Strauss, C. 1966. *The savage mind*. Chicago: University of Chicago Press.
Lévi-Strauss, C. 1969a. *The elementary structures of kinship*. London: Eyre & Spottiswoode.
Lévi-Strauss, C. 1969b. *The raw and the cooked*. New York: Harper & Row.
Lévi-Strauss, C. 1973. *From honey to ashes*. New York: Harper & Row.
Lévi-Strauss, C. 1978. *The origin of table manners*. London: Jonathan Cape.
Lomborg, B. 2001. *The skeptical environmentalist: Measuring the real state of the world*. Cambridge: Cambridge University Press.
Low, N. and B. Gleeson 1998. *Justice, society and nature: An exploration of political ecology*. London: Routledge.
Lyon, P., ed. 1974. *Native South Americans: Ethnology of the least known continent*. Boston: Little, Brown & Co.
Marcos, J. G. 1977/1978. Cruising to Acapulco and back with the Thorny Oyster set: A model for a lineal exchange system. *Journal of the Steward Anthropological Society* 9:99–132.
Marcus, G. E. and M. M. J. Fischer 1986. *Anthropology as cultural critique: An experimental moment in the human sciences*. Chicago: University of Chicago Press.
Marsden, B. and C. Smith 2005. *Engineering empires: A cultural history of technology in nineteenth-century Britain*. Houndmills: Palgrave.
Martinez-Alier, J. 1987. *Ecological economics: Energy, environment and society*. Oxford: Blackwell.
Martinez-Alier, J. 2002. *The environmentalism of the poor: A study of ecological conflicts and valuation*. Cheltenham: Edward Elgar.
Martinez-Alier, J. 2007. Marxism, social metabolism, and international trade. In *Rethinking environmental history: World-system history and global environmental change*. A. Hornborg, J. R. McNeill, and J. Martinez-Alier, eds. Lanham: AltaMira Press, pp. 221–237.
Martinez-Alier, J. 2010. Social metabolism and ecological distribution conflicts. In *International trade and environmental justice: Toward a global political ecology*. A. Hornborg and A.K. Jorgenson, eds. New York: Nova Science Publishers, pp. 1–17.
Marx, K. 1867–1893. *Capital*, vol. 1–3. London: Penguin.
Maturana, H. R. and F. J. Varela 1992. *The tree of knowledge: The biological roots of human understanding*. Boston: Shambhala.
McAnany, P. A. and N. Yoffee, eds. 2010. *Questioning collapse: Human resilience, ecological vulnerability, and the aftermath of empire*. Cambridge: Cambridge University Press.
McDonald, F. and G. McWhiney 1980. The south from self-sufficiency to peonage: An interpretation. *The American Historical Review* 85 (5):1095–1118.
McNeill, J. R. 2000. *Something new under the sun: An environmental history of the twentieth-century world*. New York: W.W. Norton & Co.
McNeill, J. R. 2010. Sustainable survival. In *Questioning collapse: Human resilience, ecological vulnerability, and the aftermath of empire*. P. A. McAnany and N. Yoffee, eds. Cambridge: Cambridge University Press, pp. 355–366.

Menget, P. 1981. From forest to mouth: Reflections on the Txicáo theory of substance. In *Food taboos in lowland South America*. K. M. Kensinger and W. H. Kracke, eds. Working Papers on South American Indians 3. Bennington: Bennington College, pp. 1–17.

Merchant, C. 1989. *Ecological revolutions: Nature, gender, and science in New England*. Chapel Hill: University of North Carolina Press.

Millennium Ecosystem Assessment 2005. *Living beyond our means: Natural assets and human well-being*. [MAweb.org]

Miller, D. 1987. *Material culture and mass consumption*. Oxford: Blackwell.

Miller, D., ed. 2005. *Materiality*. Durham: Duke University Press.

Mintz, S. 1985. *Sweetness and power: The place of sugar in modern history*. London: Penguin.

Mitchell, B. R. 1962. *Abstract of British historical statistics*. Cambridge: Cambridge University Press.

Mitchell, W. J. T. 2005. *What do pictures want? The lives and loves of images*. Chicago: University of Chicago Press.

Moore, J.W. 2000. Environmental crises and the metabolic rift in world-historical perspective. *Organization & Environment* 13(2):123–157.

Moore, J. W. 2001. Review of Sing C. Chew, 'World ecological degradation: Accumulation, urbanization, and deforestation 3000 BC–AD 2000'. *Journal of World-Systems Research* 7(2):313–316.

Moore, J. W. 2003. Capitalism as world-ecology: Braudel and Marx on environmental history. *Organization & Environment* 16(4):431–458.

Moore, J. W. 2007. Silver, ecology, and the origins of the modern world, 1450–1640. In *Rethinking environmental history: World-system history and global environmental change*. A. Hornborg, J. R. McNeill, and J. Martinez-Alier, eds. Lanham: AltaMira Press, pp. 123–142.

Moore, J. W. 2009. 'The wildness is taken from the forest': Iron, silver, and the European forest in the rise of capitalism, 1450–1800. Paper presented at the First World Congress of Environmental History, Copenhagen.

Moran, E. F., ed. 1990. *The ecosystem approach in anthropology: From concept to practice*. Ann Arbor: University of Michigan Press.

Moran, E. F. 1993. *Through Amazonian eyes: The human ecology of Amazonian populations*. Iowa City: University of Iowa Press.

Morris, C. 1978. The archaeological study of Andean exchange systems. In *Social archaeology: Beyond subsistence and dating*. C. L. Redman, ed. New York: Academic Press, pp. 315–327.

Morris, C. 1979. Maize beer in the economics, politics and religion of the Inca Empire. In *Fermented food beverages in nutrition*. C. F. Gastineau, W. J. Darby and T. B. Turner, eds. New York: Academic Press, pp. 21–34.

Munn, N. 1986. *The fame of Gawa*. Cambridge: Cambridge University Press.

Muradian, R. and S. Giljum 2007. Physical trade flows of pollution-intensive products: Historical trends in Europe and the world. In *Rethinking environmental history: World-system history and global environmental change*. A. Hornborg, J. R. McNeill, and J. Martinez-Alier, eds. Lanham: AltaMira Press, pp. 307–325.

Murra, J. V. 1962. Cloth and its functions in the Inca state. *American Anthropologist* 64:710–728.

Murra, J. V. 1975. El tráfico de mullu en la costa del Pacífico. In *Formaciones económicas y políticas del mundo andino*. Lima: Instituto de Estudios Peruanos, pp. 255–267.

Myers, T. 1983. Redes de intercambio tempranas en la hoya Amazónica. *Amazonía Peruana* 4(8):61–75.

Nadasdy, P. 2007. Adaptive co-management and the gospel of resilience. In *Adaptive co-management: Collaboration, learning, and multi-level governance*. F. Berkes, D. Armitage, and N. Doubleday, eds. Seattle: University of Washington Press, pp. 208–226.

Needham, J. 1965. *Science and civilization in China*, vol. 4, part 2. Cambridge: Cambridge University Press.

Nimuendajú, C. 1974 [1946]. Farming among the Eastern Timbira. In *Native South Americans: Ethnology of the least known continent*. P. Lyon, ed. Boston: Little, Brown, pp. 111–119.

Nordlund, C. 2010. *Social ecography: International trade, network analysis and an Emmanuelian conceptualization of ecological unequal exchange*. PhD dissertation. Lund Studies in Human Ecology 11. Lund University.

Norgaard, R. B. 1994. *Development betrayed: The end of progress and a coevolutionary revisioning of the future*. London: Routledge.

North, P. 2007. *Money and liberation: The micropolitics of alternative currency movements*. Minneapolis: University of Minnesota Press.

Nöth, W. 1998. Ecosemiotics. *Sign Systems Studies* 26:332–343.

Oberem, U. 1974 [1967]. Trade and trade goods in the Ecuadorian montaña. In *Native South Americans: Ethnology of the least known continent*. P. Lyon, ed. Boston: Little, Brown, pp. 346–357.

O'Connor, M., ed. 1994. Is capitalism sustainable? Political economy and the politics of ecology. New York: The Guilford Press.

Odum, E. P. 1989. *Ecology and our endangered life-support systems*. Sunderland: Sinauer.

Odum, H. T. 1971. *Environment, power, and society*. Hoboken: Wiley-Interscience.

Odum, H. T. 1988. Self-organization, transformity, and information. *Science* 242:1132–1139.

Odum, H. T. and J. E. Arding 1991. Emergy analysis of shrimp mariculture in Ecuador. Working paper, Coastal Resources Center, University of Rhode Island, Narragansett.

Overing, J. 1993. Death and the loss of civilized predation among the Piaroa of the Orinoco Basin. *L'Homme* 126/128:191–211.

Painter, M. and W. H. Durham, eds. 1995. *The social causes of environmental destruction in Latin America*. Ann Arbor: University of Michigan Press.

Park, R. E. 1936. Human ecology. *American Journal of Sociology* 42:1–15.

Patzek, T. W. and D. Pimentel, 2005. Thermodynamics of energy production from biomass. *Critical Reviews in Plant Sciences* 24:327–364.

Paulsen, A. C. 1974. The Thorny Oyster and the voice of God: *Spondylus* and *Strombus* in Andean prehistory. *American Antiquity* 39:597–607.

Paulsen, A. C. 1977. Patterns of maritime trade between south coastal Ecuador and western Mesoamerica, 1500 BC–AD 600. In *The sea in the Pre-Columbian world.* E. P. Benson, ed. Washington: Dumbarton Oaks Research Library and Collection, pp. 141–160.

Paulson, S. and L. L. Gezon, eds. 2005. *Political ecology across spaces, scales, and social groups*. Chapel Hill: Rutgers University Press.

Peet, R. and M. Watts, eds. 1996. *Liberation ecologies: Environment, development, social movements*. London: Routledge.

Peirce, C. S. 1931–1958. *Collected papers*, vol. 1–6. Cambridge, Mass.: Harvard University Press.

Pérez Rincón, M. 2006. Colombian international trade from a physical perspective: Towards an ecological 'Prebisch Thesis'. *Ecological Economics* 59:519–529.

Pietz, W. 1985–1988. The problem of the fetish, I–III. *Res* 9:5–17, 13:23–45, 16:105–23.

Pietz, W. 1993. Fetishism and materialism: The limits of theory in Marx. In *Fetishism as cultural discourse*. E. Apter and W. Pietz, eds. Ithaca: Cornell University Press, pp. 119–151.

Pietz, W. 1998. Afterword: How to grow oranges in Norway. In *Border fetishisms: Material objects in unstable spaces*. P. Spyer, ed. London: Routledge, pp. 245–251.

Pimentel, D. 1991. Ethanol fuels: Energy security, economics, and the environment. *Journal of Agricultural and Environmental Ethics* 4:1–13.

Pimentel, D. 2001. Ethanol fuels: Energy, economics and environmental impact. *International Sugar Journal* 103:491.

Pimentel, D., T. W. Patzek, and G. Cecil 2007. Ethanol production: Energy, economic, and environmental losses. *Reviews of Environmental Contamination and Toxicology* 189:25–41.

Pomeranz, K. 2000. *The great divergence: China, Europe, and the making of the modern world economy*. Princeton: Princeton University Press.

Pomeranz, K. and S. Topik 1999. *The world that trade created: Society, culture, and the world economy 1400 to the present*. Armonk: M.E. Sharpe.

Ponting, C. 1991. *A green history of the world: The environment and the collapse of great civilizations*. Harmondsworth: Penguin.

Porro, A. 1994. Social organization and political power in the Amazon floodplain: The ethnohistorical sources. In *Amazonian Indians from prehistory to the present: Anthropological perspectives*. A. C. Roosevelt, ed. Tucson: University of Arizona Press, pp. 79–94.

Posey, D. A. 1983. Indigenous ecological knowledge and development of the Amazon. In *The dilemma of Amazonian development*. E. F. Moran, ed. Boulder: Westview Press, pp. 135–144.

Posey, D. A. 1985. Indigenous management of tropical forest ecosystems: The case of the Kayapó Indians of the Brazilian Amazon. *Agroforestry Systems* 3(2):139–158.

Posey, D. A. and W. Balée, eds. 1989. *Resource management in Amazonia: Indigenous and folk strategies*. Advances in Economic Botany 7. Bronx: New York Botanical Garden.

Prebisch, R. 1950. *The economic development of Latin America and its principal problems*. New York: United Nations.

Radkau, J. 2008. *Nature and power: A global history of the environment*. Cambridge: Cambridge University Press.

Rappaport, R. A. 1968. *Pigs for the ancestors: Ritual in the ecology of a New Guinea people*. New Haven: Yale University Press.

Rappaport, R. A. 1979. *Ecology, meaning, and religion*. Berkeley: North Atlantic Books.

Rappaport, R. A. 1993. Humanity's evolution and anthropology's future. In *Assessing cultural anthropology*. R. Borofsky, ed. New York: McGraw-Hill, pp. 153–166.

Rathje, W. L. 1973. Classic Maya development and denouement: A research design. In *The Classic Maya collapse*. T. P. Culbert, ed. Albuquerque: University of New Mexico Press, pp. 405–454.

Redman, C. L. 1999. *Human impact on ancient environments*. Tucson: University of Arizona Press.

Renfrew, C. 1978. Trajectory discontinuity and morphogenesis: The implications for catastrophe theory for archaeology. *American Antiquity* 43:203–222.

Rice, J. 2007. Ecological unequal exchange: Consumption, equity, and unsustainable structural relationships within the global economy. *International Journal of Comparative Sociology* 48(1):43–72.

Richards, J. F. 2003. *The unending frontier: An environmental history of the early modern world*. Berkeley: University of California Press.

Rival, L. 1996. Domestication as a historical and symbolic process: Wild gardens and cultivated forests in the Ecuadorian Amazon. In *Advances in historical ecology*. W. Balée, ed. New York: Columbia University Press, pp. 232–250.

Rivière, P. G. 1969. *Marriage among the Trio: A principle of social organization*. Oxford: Oxford University Press.

Roberts, J. T. and B. C. Parks 2007. *A climate of injustice: Global inequality, North-South politics, and climate change*. Cambridge, Mass.: MIT Press.

Roberts, J. T. and B. C. Parks 2009. Ecologically unequal exchange, ecological debt, and climate justice. *International Journal of Comparative Sociology* 50(3–4):385–409.

Roosevelt, A. C. 1993. The rise and fall of the Amazon chiefdoms. *L'Homme* 126/128:255–282.

Roosevelt, A. C. 1994. Amazonian anthropology: Strategy for a new synthesis. In *Amazonian Indians from prehistory to the present: Anthropological perspectives*. A. C. Roosevelt, ed. Tucson: University of Arizona Press, pp. 1–29.

Ross, E. B. 1978. Food taboos, diet and hunting strategy: The adaptation to animals in Amazon cultural ecology. *Current Anthropology* 19(1):1–36.

Rostworowski de Diez Canseco, M. 1977. Coastal fishermen, merchants, and artisans in Pre-Hispanic Peru. In *The sea in the Pre-Columbian world*. E. P. Benson, ed. Washington: Dumbarton Oaks Research Library and Collection, pp. 167–186.

Roth, W. E. 1974 [1924/1929]. Trade and barter among the Guiana Indians. In *Native South Americans: Ethnology of the least known continent*. P. Lyon, ed. Boston: Little, Brown, pp. 159–165.

Ruyle, E. E. 1977. Energy and culture. In *The concept and dynamics of culture*. B. Banardi, ed. Paris: Mouton, pp. 209–237.

Sahlins, M. D. 1976. *Culture and practical reason*. Chicago: University of Chicago Press.

Sahlins, M. D. and E. R. Service, eds. 1960. *Evolution and culture*. Ann Arbor: University of Michigan Press.

Sale, K. 1995. *Rebels against the future: The Luddites and their war on the Industrial Revolution*. New York: Perseus.

Salomon, F. 1986. *Native lords of Quito in the age of the Incas: The political economy of north Andean chiefdoms*. Cambridge: Cambridge University Press.

Santos-Granero, F., ed. 2009. *The occult life of things: Native Amazonian theories of materiality and personhood*. Tucson: University of Arizona Press.

Schlote, W. 1952. *British overseas trade from 1700 to the 1930s*. Oxford: Basil Blackwell.

Schneider, F., G. Kallis, and J. Martinez-Alier 2010. Crisis or opportunity? Economic degrowth for social equity and ecological sustainability. *Journal of Cleaner Production* 18:511–518.

Schrödinger, E. 1944. *What is life? Mind and matter*. Cambridge: Cambridge University Press.

Schumpeter, E. B. 1960. *English overseas trade statistics 1697–1808*. Broadbridge: Clarendon Press.

Shimada, I. 1985. Perception, procurement, and management of resources: Archaeological perspective. In *Andean ecology and civilization: An interdisciplinary perspective on Andean ecological complementarity*. S. Masuda, I. Shimada and C. Morris, eds. Tokyo: University of Tokyo Press, pp. 357–399.

Shimada, I. 1987. Horizontal and vertical dimensions of prehistoric states in north Peru. In *The origins and development of the Andean state*. J. Haas, S. Pozorski and T. Pozorski, eds. Cambridge: Cambridge University Press, pp. 130–144.

Shweder, R. A. and E. J. Bourne 1984. Does the concept of the person vary cross-culturally? In *Culture theory: Essays on mind, self, and emotion*. R. A. Shweder and R. A. LeVine, eds. Cambridge: Cambridge University Press, pp. 158–199.

Simmons, I. G. 1993a. *Interpreting nature: Cultural constructions of the environment*. London: Routledge.

Simmons, I. G. 1993b. *Environmental history: A concise introduction*. Oxford: Blackwell.

Simmons, I. G. 2008. *Global environmental history 10,000 BC to AD 2000*. Edinburgh: Edinburgh University Press.

Siskind, J. 1973. *To hunt in the morning*. Oxford: Oxford University Press.

Sjödin, E. 1974. *Får*. Stockholm: LT:s förlag.

Smil, V. 1994. *Energy in world history*. Boulder: Westview.

Sponsel, L. E., ed. 1995. *Indigenous peoples and the future of Amazonia: An ecological anthropology of an endangered world*. Tucson: University of Arizona Press.

Spyer, P., ed. 1998. *Border fetishisms: Material objects in unstable spaces*. London: Routledge.

Srinivasan, U. T., S. P. Carey, E. Hallstein *et al.* 2008. The debt of nations and the distribution of ecological impacts from human activities. *Proceedings of the National Academy of Sciences* 105(5):1768–1773.

Sterling, T. 1973. *The Amazon.* Amsterdam: Time-Life International.

Summers, L. 1991. Memorandum. www.whirledbank.org/ourwords/summers.html.

Tainter, J. A. 1988. *The collapse of complex societies.* Cambridge: Cambridge University Press.

Tainter, J. A. 2006. Archaeology of overshoot and collapse. *Annual Review of Anthropology* 35:59–74.

Tambiah, S. 1968. The magical power of the word. *Man* (N.S.) 3:175–208.

Tambiah, S. 1969. Animals are good to think and good to prohibit. *Ethnology* 8:423–459.

Taussig, M. T. 1980. *The devil and commodity fetishism in South America.* Chapel Hill: University of North Carolina Press.

Taylor, K. I. 1981. Knowledge and praxis in Sanumá food prohibitions. In *Food taboos in lowland South America.* K. M. Kensinger and W. H. Kracke, eds. Working Papers on South American Indians 3. Bennington: Bennington College, pp. 24–54.

Thomas, W. L., Jr., ed. 1956. *Man's role in changing the face of the Earth.* Chicago: University of Chicago Press.

Thompson, F. M. L. 1968. The second agricultural revolution, 1815–1880. *Economic History Review* 21:62–77.

Turner, B. L., II, W. C. Clark, R. W. Kates *et al.*, eds. 1990. *The Earth as transformed by human action: Global and regional changes in the biosphere over the past 300 years.* Cambridge: Cambridge University Press.

Tuzin, D. F. 1981. Food taboos in lowland South America: A discussion. In *Food taboos in lowland South America.* K. M. Kensinger and W. H. Kracke, eds. Working Papers on South American Indians 3. Bennington: Bennington College, pp. 186–191.

Uexküll, J. von 1982 [1940]. *The theory of meaning.* Semiotica 42:25–82.

United Nations Development Program 1998. *Human development report.* United Nations Development Program.

Urban, G. 1981. The semiotics of tabooed food: Shokleng (Gé). In *Food taboos in lowland South America.* K. M. Kensinger and W. H. Kracke, eds. Working Papers on South American Indians 3. Bennington: Bennington College, pp. 76–90.

Victor, P. A. 2008. *Managing without growth: Slower by design, not disaster.* Cheltenham: Edward Elgar.

Viveiros de Castro, E. 1992. *From the enemy's point of view: Humanity and divinity in an Amazonian society.* Chicago: University of Chicago Press.

Viveiros de Castro, E. 1996. Images of nature and society in Amazonian ethnology. *Annual Review of Anthropology* 25:179–200.

Viveiros de Castro, E. 1998. Cosmological deixis and Amerindian perspectivism. *The Journal of the Royal Anthropological Institute* 4(3):469–488.

Viveiros de Castro, E. 1999. The transformation of objects into subjects in Amerindian ontologies. Paper presented at the 98th Annual Meeting of the American Anthropological Association, Chicago.

Wackernagel, M. and W. E. Rees 1996. *Our ecological footprint: Reducing human impact on the Earth.* Gabriola Island: New Society Publishers.

Wallerstein, I. 1974. *The modern world-system: Capitalist agriculture and the origins of the European world-economy in the sixteenth century.* San Diego: Academic Press.

Wallerstein, I. 1974–1989. *The modern world-system*, vol. 1–3. San Diego: Academic Press.

Wallerstein, I. 1995. Hold the tiller firm: On method and the unit of analysis. In *Civilizations and world systems: Studying world-historical change.* S. K. Sanderson, ed. Lanham: AltaMira Press, pp. 239–247.

Webb, M. C. 1973. The Petén Maya decline viewed in the perspective of state formation. In *The Classic Maya collapse.* T. P. Culbert, ed. Albuquerque: University of New Mexico Press, pp. 367–404.

Weiss, G. 1974 [1972]. Campa cosmology. In *Native South Americans: Ethnology of the least known continent.* P. Lyon, ed. Boston: Little, Brown, pp. 251–266.

Weisz, H. 2007. Combining social metabolism and input-output analyses to account for ecologically unequal trade. In *Rethinking environmental history: World-system history and global environmental change.* A. Hornborg, J. R. McNeill, and J. Martinez-Alier, eds. Lanham: AltaMira Press, pp. 289–306.

Weisz, H., F. Krausmann, C. Amann *et al.* 2006. The physical economy of the European Union: Cross-country comparison and determinants of material consumption. *Ecological Economics* 58:676–698.

White, L., Jr. 1962. *Medieval technology and social change.* Oxford: Oxford University Press.

Whitehead, N. L. 1993. Ethnic transformation and historical discontinuity in native Amazonia and Guayana, 1500–1900. *L'Homme* 126/128:285–304.

Whitehead, N. L. 1994. The ancient Amerindian polities of the Amazon, the Orinoco, and the Atlantic coast: A preliminary analysis of their passage from antiquity to extinction. In *Amazonian Indians from prehistory to the present: Anthropological perspectives.* A. C. Roosevelt, ed. Tucson: University of Arizona Press, pp. 33–53.

Widgren, M. 2007a. Precolonial landesque capital: A global perspective. In *Rethinking environmental history: World-system history and global environmental change.* A. Hornborg, J. R. McNeill, and J. Martinez-Alier, eds. Lanham: AltaMira Press, pp. 61–77.

Widgren, M. 2007b. Review of Sing C. Chew, 'The recurring Dark Ages: Ecological stress, climate changes, and system transformation'. *Environment and History* 13:503–506.

Wilk, R. R. 2007. The extractive economy: An early phase of the globalization of diet, and its environmental consequences. In *Rethinking environmental history: World-system history and global environmental change.* A. Hornborg, J. R. McNeill, and J. Martinez-Alier, eds. Lanham: AltaMira Press, pp.179–198.

Wilk, R. R. and L. C. Cliggett 2007. *Economies and cultures: Foundations of economic anthropology.* Boulder: Westview Press.

Wilkinson, R. G. 1973. *Poverty and progress: An ecological model of economic development.* London: Methuen.

Williams, M. 2003. *Deforesting the Earth: From prehistory to global crisis*. Chicago: University of Chicago Press.

Williams, M. 2007. The role of deforestation in Earth and world-system integration. In *Rethinking environmental history: World-system history and global environmental change*. A. Hornborg, J.R. McNeill, and J. Martinez-Alier, eds. Lanham: AltaMira Press, pp. 101–122.

Wilson, D. J. 1999. *Indigenous South Americans of the past and present: An ecological perspective*. Boulder: Westview Press.

Wolf, E. R. 1982. *Europe and the people without history*. Berkeley: University of California Press.

Worster, D. 1979. *Dust bowl*. Oxford: Oxford University Press.

Worster, D., ed. 1988. *The ends of the Earth: Perspectives on modern environmental history*. Cambridge: Cambridge University Press.

World Commission on Environment and Development 1987. *Our common future*. Oxford: Oxford University Press.

Wrigley, E. A. 1988. *Continuity, chance, and change: The character of the Industrial Revolution in England*. Cambridge: Cambridge University Press.

Yoffee, N. 1988. The collapse of ancient Mesopotamian states and civilization. In *The collapse of ancient states and civilizations*. N. Yoffee and G. L. Cowgill, eds. Tucson: University of Arizona Press, pp. 44–68.

Yoffee, N. and G. L. Cowgill, eds. 1988. *The collapse of ancient states and civilizations*. Tucson: University of Arizona Press.

York, R., E. A. Rosa, and T. Dietz 2003. Footprints on the Earth: The environmental consequences of modernity. *American Sociological Review* 68(2):279–300.

Index

Note: Page numbers in *italics* denote tables, those in **bold** denote figures.